Fuzzy Cooperative Games

Studies in Fuzziness and Soft Computing

Editor-in-chief
Prof. Janusz Kacprzyk
Systems Research Institute
Polish Academy of Sciences
ul. Newelska 6
01-447 Warsaw, Poland
E-mail: kacprzyk@ibspan.waw.pl
http://www.springer.de/cgi-bin/search_book.pl?series=2941

Further volumes of this series can be found at our homepage.

Vol. 49. L.I. Kuncheva
Fuzzy Classifier Design, 2000
ISBN 3-7908-1298-6

Vol. 50. F. Crestani and G. Pasi (Eds.)
Soft Computing in Information Retrieval, 2000
ISBN 3-7908-1299-4

Vol. 51. J. Fodor, B. De Baets and P. Perny (Eds.)
Preferences and Decisions under Incomplete Knowledge, 2000
ISBN 3-7908-1303-6

Vol. 52. E.E. Kerre and M. Nachtegael (Eds.)
Fuzzy Techniques in Image Processing, 2000
ISBN 3-7908-1304-4

Vol. 53. G. Bordogna and G. Pasi (Eds.)
Recent Issues on Fuzzy Databases, 2000
ISBN 3-7908-1319-2

Vol. 54. P. Sinčák and J. Vaščák (Eds.)
Quo Vadis Computational Intelligence?, 2000
ISBN 3-7908-1324-9

Vol. 55. J.N. Mordeson, D.S. Malik and S.-C. Cheng
Fuzzy Mathematics in Medicine, 2000
ISBN 3-7908-1325-7

Vol. 56. L. Polkowski, S. Tsumoto and T.Y. Lin (Eds.)
Rough Set Methods and Applications, 2000
ISBN 3-7908-1328-1

Vol. 57. V. Novák and I. Perfilieva (Eds.)
Discovering the World with Fuzzy Logic, 2001
ISBN 3-7908-1330-3

Vol. 58. D.S. Malik and J.N. Mordeson
Fuzzy Discrete Structures, 2000
ISBN 3-7908-1335-4

Vol. 59. T. Furuhashi, S. Tano and H.-A. Jacobsen (Eds.)
Deep Fusion of Computational and Symbolic Processing, 2001
ISBN 3-7908-1339-7

Vol. 60. K.J. Cios (Ed.)
Medical Data Mining and Knowledge Discovery, 2001
ISBN 3-7908-1340-0

Vol. 61. D. Driankov and A. Saffiotti (Eds.)
Fuzzy Logic Techniques for Autonomous Vehicle Navigation, 2001
ISBN 3-7908-1341-9

Vol. 62. N. Baba, L.C. Jain (Eds.)
Computational Intelligence in Games, 2001
ISBN 3-7908-1348-6

Vol. 63. O. Castillo, P. Melin
Soft Computing for Control of Non-Linear Dynamical Systems, 2001
ISBN 3-7908-1349-4

Vol. 64. I. Nishizaki, M. Sakawa
Fuzzy and Multiobjective Games for Conflict Resolution, 2001
ISBN 3-7908-1341-9

Vol. 65. E. Orłowska, A. Szalas (Eds.)
Relational Methods for Computer Science Applications, 2001
ISBN 3-7908-1365-6

Vol. 66. R.J. Howlett, L.C. Jain (Eds.)
Radial Basis Function Networks 1, 2001
ISBN 3-7908-1367-2

Vol. 67. R.J. Howlett and L.C. Jain (Eds.)
Radial Basis Function Networks 2, 2001
ISBN 3-7908-1368-0

Vol. 68. A. Kandel, M. Last and H. Bunke (Eds.)
Data Minining and Computational Intelligence, 2001
ISBN 3-7908-1371-0

Vol. 69. A. Piegat
Fuzzy Modeling and Control, 2001
ISBN 3-7908-1385-0

Vol. 70. W. Pedrycz (Ed.)
Granular Computing, 2001
ISBN 3-7908-1387-7

Vol. 71. K. Leiviskä (Ed.)
Industrial Applications of Soft Computing
ISBN 3-7908-1388-5

Milan Mareš

Fuzzy Cooperative Games

Cooperation with Vague Expectations

With 2 Figures
and 6 Tables

Physica-Verlag
A Springer-Verlag Company

Prof. Milan Mareš
Academy of Sciences of the Czech Republic
Institute of Information Theory and Automation
P.O. Box 18
18208 Prague 8
Czech Republic
Mares@utia.cas.cz

ISSN 1434-9922
ISBN 3-7908-1392-3 Physica-Verlag Heidelberg New York

Cataloging-in-Publication Data applied for
Die Deutsche Bibliothek – CIP-Einheitsaufnahme
Mareš, Milan: Fuzzy cooperative games: cooperation with vague expectations; with 6 tables / Milan Mareš.
– Heidelberg; New York: Physica-Verl., 2001
(Studies in fuzziness and soft computing; Vol. 72)
ISBN 3-7908-1392-3

Physica-Verlag Heidelberg New York
a member of BertelsmannSpringer Science+Business Media GmbH

Printed in Germany

Hardcover Design: Erich Kirchner, Heidelberg

SPIN 10797421 88/2202-5 4 3 2 1 0 – Printed on acid-free paper

To Maruška, Markéta and Martin

Acknowledgements

There are many institutions and persons whose contribution to this book cannot be omitted and overvalued.

Research the results of which are summarized in the following chapters was supported from several sources. Namely, the following projects and grants should be mentioned: The Academy of Sciences of the Czech Republic key project No. K 1075601, the Grant Agency of the Czech Republic grants No. 402/96/0414 and 402/99/0032, the Grant Agency of the Academy of Sciences of the Czech republic grant No. A 1075905, and also the project of the Ministry of Education, Youth and Sports of the Czech Republic No. VS 96063 was partly used.

Thanks are undoubtedly due to the Institute of Information Theory and Automation, Academy of Sciences of the Czech Republic, at which the research was done, and to researchers of this institute whose comments and discussions have inspired some ideas summarized below.

Special thanks deserves Professor Radko Mesiar from the Slovak Technical University in Bratislava for his friendly readiness to analyze any problem regarding fuzziness, fuzzy quantities and triangular norms, and, most of all, to Professor Dan Butnariu from the University of Haifa, Israel, for his kind goodwill to discuss mathematical points of view to economic problems. His valuable comments to an almost final version of the manuscript have significantly contributed to its content as well as to its formal level, and they have prevented the author from numerous unpleasant slips.

Professor Janusz Kacprzyk from the Polish Academy of Sciences in Warsaw deserves deep thanks for a careful editing of the final text and for his encouragement and support.

I would like to especially thank my family for a huge and long endurance and patience they had to exert during my work on this book.

Last but not least, my very deep thanks belong to my esteemed collaborator Mrs. Iva Marešová without whose accurate and patient care for the manuscript and its technical shape this book could never appear.

Preface

The present book deals with coalition games in which expected pay-offs are only vaguely known. In fact, this idea about vagueness of expectations appears to be adequate to real situations in which the coalitional bargaining anticipates a proper realization of the game with a strategic behaviour of players. The vagueness being present in the expectations of profits is modelled by means of the theory of fuzzy set and fuzzy quantities.

The fuzziness of decision-making and strategic behaviour attracts the attention of mathematicians and its particular aspects are discussed in several works. One can mention in this respect in particular the book "Fuzzy and Multiobjective Games for Conflict Resolution" by Ichiro Nishizaki and Masatoshi Sakawa (referred below as [43]) which has recently appeared in the series *Studies in Fuzziness and Soft Computing* published by Physica–Verlag in which the present book is also apperaing. That book, together with the one you carry in your hands, form in a certain sense a complementary pair. They present detailed views on two main aspects forming the core of game theory: strategic (mostly 2-person) games, and coalitional (or cooperative) games. As a pair they offer quite a wide overview of fuzzy set theoretical approaches to game theoretical models of human behaviour.

Game theory in general, and also its fuzzified modification, is characterized by a great deal of concepts and definitions. With regard to this, also this book mostly presents a descriptive survey of concepts, motivations for their introduction, and relations among them. Their definition and properties, presented and introduced in the following chapters, are more precisely derived in the referenced literature.

The body of the book is divided into four main parts and seventeen chapters. The first part is of a rather introductory character as it summarizes auxiliary concepts of fuzzy quantities theory and deterministic coalition game theory as well as methodological paradigms on which the following parts are based. The main notions of both branches of the coalition game theory – the ones of games with and without side-payments – are introduced, and fuzzy set theoretical tools for their modified processing are presented.

The next two parts constitute the main content of this book. They introduce and study models of fuzzy coalition games and their characteristics. Part II is devoted to games with side-payments and Part III deals with games without side-payments. In both cases, a fuzzy coalition game is defined as a fuzzy extension of its deterministic counterpart. Pay-offs of the deterministic games are considered for the most possible (in fact, certain) values of fuzzy pay-offs of their fuzzified counterparts. In such fuzzy game models, the main interest is focused on a relation between various elements or properties of the original deterministic games and their fuzzy extensions. In both parts of the book, the first chapters are devoted to the properties like superadditivity, subadditivity or additivity and the main subjects of further chapters are related to the concept of a core, effectivity of coalitions and to some other closely related topics. A relation of domination of imputations is especially significant for the coalition games without side-payments, and its formalism essentially influences the validity of most of results presented in Part III.

The last Part IV completes the previous ones by topics related to the main model of fuzzy cooperation. It briefly deals with some alternative approaches to aspects of the fuzzification of coalition games. Chapter 14 deals with so-called linear coalition games. They represent a link between the games with and without side-payments, and in the deterministic games both these models, when applied, lead to fully equivalent results. This equivalence is rather broken in the fuzzy extensions of the original deterministic coalition game pattern, and the differences offer a possibility to compare both approaches to the coalition games. Chapter 15 only briefly discusses a possibility to modify, namely simplify, the domination relation in the games without side-payments. As already mentioned above, this relation significantly influences the form of other components of the game, especially its core. The coalition game models treated in this book usually suppress the strategic background of the pay-off values. In the deterministic coalition games it can be, at least in some cases, reconstructed from the 2-players coalition–anticoalition games theory. A fuzzified modification of this reconstruction is briefly discussed in Chapter 16. The last Chapter 17 discusses an alternative approach to the theory of fuzzy quantities, one of the main tools for studying fuzzy coalition games. The application of this modified approach can essentially influence most of results of the theory of fuzzy coalition games with side-payments. As formal mathematical tools for it are still under preparation and development, their application to the coalition game theory could only be briefly indicated.

There are many works which are related either to games and decision-making theory or to fuzzy set theoretical tools used in this book. Some of them are included in the list of references.

Table of Contents

Part I

PRELIMINARIES

Before starting the presentation of fuzzy cooperation models, it may be worthwhile to recall some basic heuristic ideas that are in their background, and some general concepts of both fuzzy quantities and coalition game theories including the notation to be used in the following chapters.

Fuzzy coalition games investigated in the following parts are fuzzy extensions of their classic crisp counterparts, and their formal presentation is based on the concept of fuzzy quantities.

The first and the last chapters are devoted to informal heuristic discussion of the main problems treated in this book, and of the methodology used for their solution. The second chapter summarizes the main properties of fuzzy quantities and fuzzy sets as a representation of vague knowledge of expected pay-offs. The third chapter is devoted to the classic deterministic coalition game theory. Both basic models – the coalition games with and without side-payments – are briefly characterized and their basic concepts like a core, additivity-like properties, effectivity of coalitions, the Shapley value, balancedness, convexity, etc. are explained in their deterministic form.

1. Introduction

Game theory includes not only strategic conflicts but also a possibility of cooperation and coalition forming. The development of the theory of coalition games brought two principal types of cooperation that have led to games with and without side-payments. In both cases, each player tries to maximize his or her (expected) profit and joins a coalition which promises the best individual expectations. In the games with side-payments each coalition is assumed to win some total profit that is distributed among its members, and the players tend to make this total pay-off as high as possible and to negotiate their individual participation in it as large as possible. In the games without side payments the individual profits of coalition members are supposed to be given by rules of the game and by the strategy of the coalition. A possible re-distribution of pay-offs does not preserve the values of the utilities and, hence, it cannot be considered for side-payments of the former type. The coordination of behaviour of members of the coalition necessarilly respects the optimality of all the vectors of individual profits.

The classic theory of bargaining and coalition forming in both types of games is based on an assumption that the players, at the very beginning of the negotiation process, know exactly the expected outcomes. It means that they know the total pay-offs of coalitions which will be distributed in the first case, and the vectors of individual outcomes of the cooperation in the second case. In fact, the words "expected outcomes" show a certain degree of subjectivity which can differ from an objective certainty. The players' expectations can be more or less optimistic or, in the most realistic evaluation, they can point at the value of the 2-players, coalition-anticoalition, game as the guaranteed outcome for rational coalition. But, even if we interpret "expected outcome" in such a subjective way, the proper nature of input information in the negotiation process is essentially much more vague.

Usually, realistic cooperative situations are subjected to numerous unpredictable and statistically hard to describe influences which can be only vaguely incorporated in the bargaining strategies of their participants. Subjective expectations play a significant role in the practical evaluation of such phenomena like a possible behaviour of the members of anticoalition, external influences on the result of the game, changes of the "rules" (sets of admissible strategies and resulting utility functions) during the realization of the negotiated procedures, development of the individual utility functions evaluating the practical results of the game, and others. Such uncertainty (imprecision) can be modelled by fuzzy sets and fuzzy quantities theory which reflect aspects that are not incorporated in most game theoretical models of interpersonal behaviour. Namely, it is natural to describe expected outcomes as either fuzzy quantities (in the case of games with side-payments) or as fuzzy sets of multidimensional real vectors (in the case of games without side-payments).

Uncertainty regarding the expected results of cooperative behaviour can appear in many practical situations. The following examples illustrate only some of them. The formal tools introduced in the next parts are meant to contribute to the description and analysis of such situations. Here, we can mention their general ideas only and to offer a heuristic formulation of some questions to be answered by mathematical models.

Example 1.1 Let us consider two economic subjects who wish to allocate their capital in the way bringing maximal profit. They can proceed individually and then each of them can reach a profit of 20 % of their invested stock. But they can also cummulate their capitals and invest the joint means in a qualitatively new way. However, the possible profit is subject to an uncertainty. It can be at least 15 % but it can reach, with a possibility evaluated by degree 0.6 (where 1.0 is certainty and 0.0 is impossibility) even 30 % of the invested stock. Will the joint investment be made and what is the possibility of its rationality? What is the possibility that the individual investment is better, what are the rational partitions of the commonly attained profit, and what is the possibility of this?

□

The previous example assumed the possibility to redistribute the attained common profit of the coalition among its members almost in an arbitrary way. This follows from the fact that the profit is represented by an amount of money which can be arbitrarily partitioned. There are also situations in which the profit (or utility) is not separable from individual participants and in which the negotiation process regards only the coordination of behaviour of the relevant subjects.

Example 1.2 Let us consider a parliament with four parties, named A, B, C and D. Let none of them be a major party (having a majority). To be able to form an effective government, some of them are to form a coalition. The parties entering a coalition accept the fact that they have to find a compromise between their political programmes, and they attempt to maximize the percentual rate of their own programmes being included into the common programme of the formed government. A party not participating in the governing majority has no possibility to incorporate its demands into the governmental programme and then it attains the zero "utility". The utilities attained by the coalitional parties are quantitatively represented by the percentage of their political goals incorporated in the dovernmental programme.

Some phenomena are here more or less obvious. Usually, there exists a principle of a "minimal extent of coalitions" which means that a coalition that has already attained a majority does not accept further members. Moreover, the percentual influence on the governmental programme (which represents the individual utility "pay-off" of the parties) can be established only approximately with a great deal of vagueness and subjectivity. Finally, these individual "pay-offs" cannot be redistributed. Each formulation of the governmental political programme uniquely determines a degree of satisfaction of the parties in coalition.

Let us consider a parliament with the four parties mentioned above. Let us suppose, moreover, that the parliament has 200 members, and: party A has 80 votes, party B has 60 votes and parties C and D has 30 votes each. Then the possible majority coalitions are $\{A, B\}$, $\{A, C\}$, $\{A, D\}$, and $\{B, C, D\}$. From the point of view of party B, for example, there are two possible majority coalitions to participate at: either coalition $\{A, B\}$ or $\{B, C, D\}$. And, of course, B can also be in the opposition (with utility 0) if coalitions $\{A, C\}$ or $\{A, D\}$ are formed. The utility resulting from the partitions in coalitions can be, for party B, estimated as follows. In the case of partnership with A, the acceptance of 60 – 75 % of its programme is practically guaranteed and up to 85 % is relatively possible. In the case of participation in coalition $\{B, C, D\}$, the acceptance of 40 – 70 % of the programme of B can be taken for sure and there exists a non-zero possibility (but not certainty) that the accepted part of the programme will be even 90 %. This may be interpreted as that the guaranteed part of the accepted programme represents the "intersection" (i.e. practically identical parts) of the programmes and a possibly attainable part covers some "tollerance area" in which the parties are principally ready to negotiate compromises. A definitive leading coalition which will follow from the negotiations depends on the proportions of expected satisfactions of all parties. Anyway, it is possible to ask what are the expected results of political negotiations in the parliament, how to mathematically represent the degrees of expectation of these results, how to derive from them the possibility of particular results of negotiation and, maybe, some other questions regarding the expected stability or expected optimality of the coalitions.

□

This fuzzification of input data of the coalition forming negotiation has its deep consequences for its other – let us say output – components. A vague knowledge of the expected pay-offs leads to the vagueness of their analytic properties like the superadditivity or convexity, and also to the vagueness of the formed coalitions and negotiated distribution of the profits. Players having only an uncertain idea about their pay-offs resulting from the coalitional cooperation cannot certainly decide which coalition is the best one for them. If profits are expected only as some possible values, then also coalitions and coalition structures are formed with a certain possibility, only. Moreover, the fact that the cooperation increases the common profit (the superadditivity of coalitional outputs) can also be valid with some possibility, and the distribution of the common profit which is stable against objections of some sub-coalitions (the core of the game) is stable with some degree of possibility, too.

It is to be emphasized that the approach to the fuzzy cooperation, i.e. to fuzzy coalition games, presented in this work is not the only possible one. It is also possible (and it was done, e.g., in [15]) to assume that each player participates, parallelly, in more coalitions and an "intensity" of his participations generates a fuzzy set theoretical structure in the game model. This approach is very inspiring and well reflecting the properties of cooperative positions of individuals in the real world. The following chapters of this book will study another property of a vague cooperation, namely relations between vague expectations and a consequent vague behaviour of the cooperating agents. In this sense results presented in this work may well explain differences in the interest of particular agents in various coalitions.

2. Fuzzy Quantities

Fuzzy sets, introduced in the mid-1960s by Zadeh [54], have since then became well known and widely used. Here, we only briefly recall their special case investigated, for example, in [16, 25, 28], namely the fuzzy quantities.

In the whole book we denote by R the set of all real numbers. Any fuzzy subset a of R with a membership function $\mu_a : R \to [0,1]$ satisfying:

$$\text{there exists } x_a \in R \text{ such that } \mu_a(x_a) = 1; \tag{2.1}$$

$$\text{there exist } x_a^{(1)}, x_a^{(2)} \in R,\ x_a^{(1)} < x_a < x_a^{(2)}$$
$$\text{such that } \mu_a(x) = 0 \text{ for all } x \notin [x_1, x_2], \tag{2.2}$$

is called a *fuzzy quantity* and any number x_a for which (2.1) holds is called a *modal value* of a. The set of all fuzzy quantities will be denoted by $\mathbb{R}$.

It is possible to define algebraic operations over $\mathbb{R}$. The *addition* can be defined by means of the so-called *extension principle* (see [16]): If $a, b \in \mathbb{R}$, then their sum $a \oplus b$ is also a fuzzy quantity, and its membership function $\mu_{a\oplus b}$ is derived from μ_a and μ_b by

$$\mu_{a\oplus b}(x) = \sup_{y\in R} \left[\min(\mu_a(y), \mu_b(x-y))\right], \tag{2.3}$$

for any $x \in R$. More details about this operation can be found in [16, 25, 28].

Here, we briefly mention only the definition of the opposite element to $a \in \mathbb{R}$ as fuzzy quantity $-a \in \mathbb{R}$ such that

$$\mu_{-a}(x) = \mu_a(-x) \quad \text{for } x \in R. \tag{2.4}$$

Then, the *substraction* of two fuzzy quantities a and b is defined as the sum

$$a \oplus (-b).$$

It is easy to verify that for $a, b, c \in \mathbb{R}$, we have

$$a \oplus b = b \oplus a, \quad (a \oplus b) \oplus c = a \oplus (b \oplus c) \tag{2.5}$$

but $(a \oplus b) \oplus (-b)$ is not equal to a.

The linear operations, the addition and the multiplication by a crisp number, over fuzzy quantities will be the main mathematical tools for handling fuzzy coalition games with side-payments. Their analytical properties, summarized in more detailed way in , e.g., [25], are quite clear.

It is useful to introduce the notion of a fuzzy quantity degenerated into one value. Let $r \in R$, then $\langle r \rangle \in I\!R$ is a fuzzy quantity the with membership function

$$\begin{aligned} \mu_{\langle r \rangle}(x) &= 1 \quad \text{for } x = r, \\ &= 0 \quad \text{for } x \neq r. \end{aligned} \tag{2.6}$$

Analogously to the addition, it is also possible to define the multiplication of fuzzy members (see, e.g., [16, 25, 28]). In the following sections we only need the product of a crisp and a fuzzy number. Let $r \in R$ and $a \in I\!R$. Then the *crisp product* $r \cdot a$ is a fuzzy quantity with the membership function

$$\begin{aligned} \mu_{r \cdot a}(x) &= \mu_a(x/r) \quad \text{for } x \neq r, \\ &= \mu_{\langle 0 \rangle}(x) \quad \text{for } r = 0,\ x \in R. \end{aligned} \tag{2.7}$$

Formula (2.3) can also be used in case of the sum of a fuzzy number $a \in I\!R$ and a crisp number $r \in R$, $a + r = a \oplus \langle r \rangle$, i.e. for any $x \in R$ there holds

$$\mu_{a+r}(x) = \mu_a(x - r).$$

It is not difficult to verify that for $a, b \in I\!R$ and $r \in R$

$$r(a \oplus b) = r \cdot a \oplus r \cdot b \tag{2.8}$$

but for $a \in I\!R$, $r_1, r_2 \in R$, the equality between $(r_1 + r_2) \cdot a$ and $(r_1 \cdot a) \oplus (r_2 \cdot a)$ does not generally hold.

Example 2.1 Let us consider two fuzzy quantities a and b with the membership functions

$$\begin{aligned} \mu_a(x) &= x - 1 && \text{for } x \in [1, 2], \\ &= 3 - x && \text{for } x \in [2, 3], \\ &= 0 && \text{for } x \notin [1, 3], \end{aligned}$$

$$\begin{aligned} \mu_b(x) &= x - 4 && \text{for } x \in [4, 5], \\ &= 7/2 - x/2 && \text{for } x \in [5, 7], \\ &= 0 && \text{for } x \notin [4, 7] \end{aligned}$$

then it is easy to compute

$$\begin{aligned} \mu_{a \oplus b}(x) &= x/2 - 5/2 && \text{for } x \in [5, 7], \\ &= 10/3 - x/3 && \text{for } x \in [7, 10], \\ &= 0 && \text{for } x \notin [5, 10]. \end{aligned}$$

The product with crisp number, let us say $2 \cdot a$, is a fuzzy quantity with $\mu_{2\cdot a}$ such that

$$\begin{aligned}\mu_{2\cdot a}(x) &= x/2-1 \quad \text{for} \quad x\in[2,4],\\ &= 3-x/2 \quad \text{for} \quad x\in[4,6],\\ &= 0 \quad \text{for} \quad x\notin[2,6],\end{aligned}$$

and the product $b\cdot(1/3)=b/3$ has the membership function

$$\begin{aligned}\mu_{b/3}(x) &= 3x-4 \quad \text{for} \quad x\in[4/3,5/3],\\ &= 7/2-3x/2 \quad \text{for} \quad x\in[5/3,7/3],\\ &= 0 \quad \text{for} \quad x\notin[4/3,7/3].\end{aligned}$$

□

Example 2.2 For fuzzy quantities with discrete values the general outlook is not very different. Let us consider $a, b \in \mathbb{R}$ with μ_a and μ_b such that

$$\begin{aligned}&\mu_a(1)=1/3, \quad \mu_a(2)=2/3, \quad \mu_a(3)=1, \quad \mu_a(x)=0, \text{ otherwise},\\ &\mu_b(4)=1/2, \quad \mu_b(5)=1, \quad \mu_b(6)=1/2, \quad \mu_b(x)=0, \text{ otherwise}.\end{aligned}$$

Then, evidently, we have

$$\begin{aligned}&\mu_{a\oplus b}(5)=1/3, \quad \mu_{a\oplus b}(6)=1/2, \quad \mu_{a\oplus b}(7)=1,\\ &\mu_{a\oplus b}(8)=1, \quad \mu_{a\oplus b}(9)=1/2, \quad \mu_{a\oplus b}(x)=0, \text{ otherwise}.\end{aligned}$$

Even the product with a crisp number can easily be constructed. For example, $2\cdot a$ is defined as

$$\mu_{2\cdot a}(2)=1/3, \quad \mu_{2\cdot a}(4)=2/3, \quad \mu_{2\cdot a}(6)=1, \quad \mu_{2\cdot a}(x)=0, \text{ otherwise},$$

and for $b/3$

$$\mu_{b/3}(4/3)=1/2, \quad \mu_{b/3}(5/3)=1, \quad \mu_{b/3}(2)=1/2, \quad \mu_{b/3}(x)=0, \text{ otherwise}.$$

□

Some of the algebraic properties of the operations defined above were briefly formulated in (2.5) and (2.7). The validity of some other properties of a linear space for $\mathbb{R}$ as, e.g.,

$$1\cdot a = a \quad \text{or} \quad (r_1\cdot r_2)\cdot a = r_1\cdot(r_2\cdot a), \qquad r_1, r_2 \in R, \ a\in\mathbb{R},$$

is evident.

Some other algebraic properties are however valid only in a weakened form which follows from a specific character of the fuzzy zero (cf. [25, 28]).

Let us remcall now a few important concepts.

If $a \in \mathbb{R}$, then we say that a is 0-symmetric iff

$$\mu_a(x) = \mu_a(-x) \quad \text{for all} \quad x \in R.$$

If $a, b \in \mathbb{R}$, then we say that they are additively equivalent and write $a \sim_\oplus b$ iff there exist 0-symmetric fuzzy quantities s_1 and s_2 such that

$$a \oplus s_1 = b \oplus s_2.$$

The set of 0-symmetric fuzzy quantities is interpreted as the fuzzy zero in the class of sets of additively equivalent fuzzy quantities. In this sense some other linearity properties are fulfilled. Namely, if $a, s \in \mathbb{R}$ and s is 0-symmetric, then

$$a \oplus s \sim_\oplus a, \quad a + (-a) \sim_\oplus s$$

(where (2.4) was used).

Of course, properties (2.5) and (2.8) are also valid as the additive equivalencies, as well as $1 \cdot a \sim_\oplus a$ and $(r_1 \cdot r_2) \cdot a \sim_\oplus r_1 \cdot (r_2 \cdot a)$. Moreover, we say that $a \in \mathbb{R}$ is *trapezoidal* iff there exist $x_1^{(a)} < x_2^{(a)} \le x_3^{(a)} < x_4^{(a)}$ such that for, all $x \in R$,

$$\begin{aligned}
\mu_a(x) &= \frac{x - x_1^{(a)}}{x_2^{(a)} - x_1^{(a)}} && \text{for} \quad x \in [x_1^{(a)}, x_2^{(a)}], \\
&= 1 && \text{for} \quad x \in [x_2^{(a)}, x_3^{(a)}], \\
&= \frac{x_4^{(a)} - x}{x_4^{(a)} - x_3^{(a)}} && \text{for} \quad x \in [x_3^{(a)}, x_4^{(a)}], \\
&= 0 && \text{for} \quad x \notin [x_1^{(a)}, x_4^{(a)}],
\end{aligned}$$

and we say that $b \in \mathbb{R}$ is *almost trapezoidal* iff there exist a trapezoidal $a \in \mathbb{R}$ and a 0-symmetric $s \in R$ such that

$$b = a \oplus s.$$

It is not difficult to prove[1] that for arbitrary $r_1, r_2 \in R$ and an almost trapezoidal $a \in \mathbb{R}$ there holds

$$(r_1 + r_2) \cdot a \sim_\oplus r_1 \cdot a \oplus r_2 \cdot a.$$

Fuzzy quantities can be compared and ordered. There exist several definitions of the ordering relation over $\mathbb{R}$. Some of them are mentioned in [16] or briefly summarized in [28], some are mentioned in [18] and in [53]. Too many

[1] The proof of this statement is limited to almost trapezoidal (and trapezoidal or triangular as their special case) fuzzy quantities. Nevertheless, there exists an unproven but very probable hypothesis that the validity of the below equivalence is much wider, namely, that it is valid for an arbitrary fuzzy quantity.

definitions of the ordering relation indicate a certain hesitation regarding the very nature of comparison between vague quantities. Evidently, each of those relations corresponds to specific needs in particular applications. In this work we choose the one introduced in [16] and discussed also in [28] that is based on an idea that the validity of some relation between quantities with vague values is also generally vague. It means that the ordering relation between fuzzy quantities is to be a fuzzy relation which is, for every pair $a, b \in I\!R$, valid with some possibility. In symbols, for any pair of fuzzy quantities $a, b \in I\!R$ the possibility of the relation $a \succsim b$ is defined as a real number $\nu_{\succsim}(a,b) \in [0,1]$ such that

$$\nu_{\succsim}(a,b) = \sup_{\substack{x,y \in R \\ x \geq y}} [\min(\mu_a(x), \mu_b(y))] . \tag{2.9}$$

Analogously, the equivalence, or "similarity", relation between $a, b \in I\!R$, denoted by $a \sim b$, is a fuzzy relation, and it is valid with the possibility

$$\nu_{\sim}(a,b) = \sup_{x \in R} [\min(\mu_a(x), \mu_b(x))] . \tag{2.10}$$

Some properties of these relations are mentioned in [28] and [16]. Let us remember here that if $\nu_{\succsim}(a,b) = \nu_{\succsim}(b,a) = 1$, then also $\nu_{\sim}(a,b) = 1$ and, generally, $\nu_{\sim}(a,b) \geq \min(\nu_{\succsim}(a,b), \nu_{\succsim}(b,a))$, where the equality holds if both $\mu_a(x)$ and $\mu_b(y)$ are increasing for $x < x_a$ and $y < y_b$, and decreasing for $x > x_a$ and $y > y_b$, respectively (with x_a and y_b being modal values of a and b).

Relations (2.9) and (2.10) can also be used for the comparison of fuzzy and crisp numbers. Namely, if $a \in I\!R$ and $r \in R$, then the possibility of $a \succsim r$ is equal to the possibility of $a \succsim \langle r \rangle$, i.e.

$$\nu_{\succsim}(a,r) = \sup\left(\mu_a(x) : x \in R, x \geq r\right), \tag{2.11}$$

and the possibility of $r \succsim a$ is, obviously, equal to

$$\nu_{\succsim}(r,a) = \sup\left(\mu_a(x) : x \in R, x \leq r\right). \tag{2.12}$$

Relation (2.10) implies an obvious coincidence between our fuzzy ordering and the basic definition of the membership function

$$\nu_{\sim}(r,a) = \nu_{\sim}(a,r) = \mu_a(r). \tag{2.13}$$

Let $\boldsymbol{A}$ be a set of fuzzy quantities. Then it is possible to construct its subset $\boldsymbol{A}_{\max} \subset \boldsymbol{A}$ of maximal elements of $\boldsymbol{A}$ due to relation (2.9). Since the ordering of fuzzy quantities $\succsim$ is a fuzzy relation valid with some possibilities, then the set $\boldsymbol{A}_{\max}$ is also a fuzzy subset of $\boldsymbol{A}$ where each $a \in \boldsymbol{A}$ belongs to $\boldsymbol{A}_{\max}$ with a possibility $\nu_{\max} : \boldsymbol{A} \to [0,1]$. The set $\boldsymbol{A}_{\max} \subset \boldsymbol{A}$ contains, heuristically, the quantities from $\boldsymbol{A}$ which are "greater" (in the sense of $\succsim$) than any other quantity from $\boldsymbol{A}$. In the notation introduced above, we have

$$\nu_{\max}(a) = \min_{b \in \boldsymbol{A}} \nu_{\succsim}(a, b). \tag{2.14}$$

Analogously, the set $\boldsymbol{A}_{\min}$ of minimal elements from $\boldsymbol{A}$ is also a fuzzy subset of $\boldsymbol{A}$ with the membership function $\nu_{\min} : \boldsymbol{A} \to [0, 1]$. The value $\nu_{\min}(a)$ for $a \in \boldsymbol{A}$ determines the possibility that a is "smaller" than any other quantity from $\boldsymbol{A}$. Formally, we have

$$\nu_{\min}(a) = \min_{b \in \boldsymbol{A}} \nu_{\succsim}(b, a). \tag{2.15}$$

There exist several different definitions of maximal or minimal fuzzy quantities, depending on the form of the ordering relation. Their adequacy depends on the actual situation and application. For our purpose we will use the ones shown above.

The above ordering relation has a quite a good interpretation and looks natural. The following examples illustrate its elementary properties.

Example 2.3 Let us consider two fuzzy quantities $a, b \in \mathbb{R}$ with the membership functions

$$\begin{aligned} \mu_a(x) &= x - 3 && \text{for} \quad x \in [3, 4], \\ &= 3 - x/2 && \text{for} \quad x \in [4, 6], \\ &= 0 && \text{otherwise,} \end{aligned}$$

$$\begin{aligned} \mu_b(x) &= x - 5 && \text{for} \quad x \in [5, 6], \\ &= 7 - x && \text{for} \quad x \in [6, 7], \\ &= 0 && \text{otherwise.} \end{aligned}$$

Then, $\nu_{\succsim}(b, a) = 1$, and $\nu_{\succsim}(a, b) = 1/3$.

□

Example 2.4 Let us consider three fuzzy quantities with integer values a, b, c with the membership functions

$$\begin{array}{lll} \mu_a(1) = 0.9, & \mu_a(3) = 1, & \mu_a(5) = 0.8, \\ \mu_b(2) = 0.6, & \mu_b(3) = 0.7, & \mu_b(4) = 1, \\ \mu_c(5) = 1, & \mu_c(6) = 0.9, & \end{array}$$

such that $\mu_a(x)$, $\mu_b(x)$ and $\mu_c(x)$ vanish in other cases.

Then it is not difficult to verify that

$$\begin{array}{ll} \nu_{\succsim}(a, b) = 0.8, & \nu_{\succsim}(b, a) = 1.0, \\ \nu_{\succsim}(a, c) = 0.8, & \nu_{\succsim}(c, a) = 1.0, \\ \nu_{\succsim}(b, c) = 0, & \nu_{\succsim}(c, b) = 1.0, \end{array}$$

and

$$\begin{array}{lll} \nu_{\max}(a) = 0.8, & \nu_{\max}(b) = 0, & \nu_{\max}(c) = 1, \\ \nu_{\min}(a) = 1, & \nu_{\min}(b) = 0.8, & \nu_{\min}(c) = 0. \end{array}$$

□

The above concepts and properties related to fuzzy quantities will be used mainly for the modelling of vagueness in games with side-payments.

3. Deterministic Coalition Games

The theory of coalition games represents a standard mathematical model of cooperation among agents whose aim is to increase their individual profits. The theory is deeply developed and includes numerous concepts and results. Here, we recall its elementary parts that will be fuzzified in the following chapters.

In the whole work we denote by I a non-empty and finite *set of players*. Each subset K of I is called a *coalition* and any partition $\mathcal{K}$ of I into disjoint coalitions, $\mathcal{K} = (K_1, \dots, K_m)$, such that

$$K_i \cap K_j \neq \emptyset, \quad i \neq j, \ i, j = 1, \dots, m, \quad K_1 \cup \dots \cup K_m = I, \tag{3.1}$$

is called a *coalition structure.*

It is usual to distinguish two main types of coalition games which differ in the possibility of distribution of coalitional incomes among the players.

3.1 Coalition Games With Side–Payments

The basic and formally simpler model of coalition game is based on the assumption of the existence of a universal and linear representative of utility which can be used for the distribution of the coalition's total profit among its members without any deformation of the utility value (cf. [46, 48, 19]). This means that, due to this model, each coalition winns some utility which can be distributed among its members with respect to a coalitional agreement, and the sum of utilities obtained by the coalition members preserves the value of a distributed common income.

With respect to this principle, these games are sometimes also called *transferable utility games.* In this work we preserve the more traditional term, i.e. coalition games.

Formally, a *coalition game with side-payments* is defined as a pair (I, v) where I is a set of players and $v : 2^I \to R$ is a mapping called a *characteristic function* of the game that connects each coalition $K \subset I$ with a real number

$v(K) \in R$ representing the total profit of the coalition K. We assume that for the empty coalition, $\emptyset \subset I$, $v(\emptyset) = 0$.

The distribution of profit among the players is represented by real valued vectors $\boldsymbol{x} = (x_i)_{i \in I} \in R^I$, called *imputations*. Evidently, an imputation $\boldsymbol{x}$ is *admissible for a coalition* K iff

$$\sum_K x_i \leq v(K).$$

If $\sum_K x_i \geq v(K)$, then the coalition K cannot negotiate any imputation $\boldsymbol{y} = (y_i)_{i \in K}$ which is uniformly better than $\boldsymbol{x}$ for any player in K, i.e. for which $y_i > x_i$, for all $i \in K$.

We say that a coalition game (I, v) is *superadditive* if for any pair of disjoint coalitions $K, L \subset I$, $K \cap L = \emptyset$, there holds

$$v(K \cup L) \geq v(K) + v(L). \tag{3.2}$$

Analogously, we say that is *subadditive* iff for any $K, L \subset I$, $K \cap L = \emptyset$, there holds

$$v(K \cup L) \leq v(K) + v(L). \tag{3.3}$$

We say that a game (I, v) is *additive* iff it is both, superadditive and subadditive, i.e. iff for any $K, L \subset I$, $K \cap L = \emptyset$, there is

$$v(K \cup L) = v(K) + v(L). \tag{3.4}$$

Stable agreements in a coalition game (I, v), i.e. agreements about the distribution of profits among the players which are possible in the game and which cannot be broken by some better offer of any coalition, are represented by the concept of a core.

More precisely, let us introduce a domination relation between imputations via a coalition as follows: $\boldsymbol{x}$ dominates $\boldsymbol{y}$ via K, $K \subset I$, $\boldsymbol{x}, \boldsymbol{y} \in R^I$, iff $x_i \geq y_i$ for all $i \in K$, and $y_j > y_j$ for some $j \in K$. Then, evidently, $\boldsymbol{x} \in R^I$ is accessible for the coalition K iff

$$\sum_K \boldsymbol{x}_i \leq v(K)$$

and $\boldsymbol{y}$ cannot be dominated by any $\boldsymbol{x} \in R^I$ via K iff

$$\sum_K y_i \geq v(K).$$

This means that the *core* of a game (I, v) is the set of imputations $C \subset R^I$ such that

$$\begin{aligned} C &= \{x \in R^I : \text{ there exists a coalition structure } \mathcal{K} \\ &\quad \text{such that } \sum\nolimits_K x_i \le v(K) \\ &\quad \text{for all } K \in \mathcal{K} \text{ and, moreover,} \\ &\quad \sum\nolimits_L x_i \ge v(L) \text{ for all } L \subset I\}. \end{aligned} \tag{3.5}$$

If a game (I, v) is superadditive then, obviously,

$$\begin{aligned} C &= \{x \in R^I : \sum\nolimits_I x_i \le v(I)\} \\ &\quad \text{and } \sum\nolimits_L x_i \ge v(L) \text{ for all } L \subset I\}, \end{aligned} \tag{3.6}$$

as $v(I) \ge \sum_{K \in \mathcal{K}} v(K)$ for any coalition structure.

A special form (3.6) of the definition of a core is more frequently used but the superadditivity is neither a necessary nor a sufficient condition for the non-emptiness of a core of this type. For this purpose, the following concept is introduced.

We say that a coalition game (I, v) is *convex* iff for every pair of coalitions $K, L \subset I$, there holds

$$v(K \cup L) + v(K \cap L) \ge v(K) + v(L). \tag{3.7}$$

It can easily be seen that convexity implies superadditivity. A well-known result of the coalition game theory (cf. [46, 19]) shows that convexity is a sufficient condition for the existence of a non-empty core (3.6) in the case when a core is generated by the coalition of all players.

The necessary and sufficient condition for the existence of a non-empty core (3.6) needs the introduction of an auxiliary concept.

We say that a set of coalitions $\mathcal{L} = \{L_1, \ldots, L_m\}$ is *balanced* if there exist positive real numbers $\alpha_1, \ldots, \alpha_m$, $\alpha_j > 0$, $j = 1, \ldots, m$, such that for every player $i \in I$, there is

$$\sum\nolimits_{j=1, i \in L_j}^{m} \alpha_j = 1. \tag{3.8}$$

Then, (I, v) is a *balanced coalition game* iff for every balanced set of coalitions $\mathcal{L} = \{L_1, \ldots, L_m\}$ with coefficients $\alpha_1, \ldots, \alpha_m$ fulfilling (3.8) we also have

$$\sum\nolimits_{j=1}^{m} \alpha_j \cdot v(L_j) \le v(I). \tag{3.9}$$

The balancedness of a coalition game is a necessary and sufficient condition for the existence of a non-empty core which is realized in the coalition of all players, as shown in [10, 46, 19] and other works.

There obviously exist games (e.g., subadditive ones) whose core is non-empty, realized by some coalition structure $\mathcal{K}$ not containing the universal coalition I, and which are not balanced (see, e.g., Example 3.4). A balanced game need not be superadditive regarding some "smaller" coalitions. But its core is realized by the coalition I and, consequently, it takes the form of (3.6).

The concept of a core represents stable imputations as a result of bargaining. It is also possible to ask which coalitions may appear in the process of bargaining, at least as some temporary result of negotiations.

We say that $K \subset I$ is an *effective coalition* iff there exists an imputation $\boldsymbol{x}_K = (x_i)_{i \in K} \in R^K$ which is accessible for K, i.e.

$$\sum_{i \in K} x_i \leq v(K)$$

and which cannot be effectively protested by any subcoalition $J \subset K$, i.e.

$$\sum_J x_i \geq v(J) \quad \text{for all} \quad J \subset K.$$

There exists an obvious relation between a core and effectiveness following from the essence of their construction. Namely, a coalition I is effective if and only if the core C fulfilling (3.6) is non-empty. Nevertheless, the concept of a coalitional effectiveness deals much more with particular coalitions and the possibility of their formation during the bargaining process.

Coalition games with side-payments are usually quite clear structures, at least if the set of players is not very large.

Example 3.1 Let us consider a 3-players coalition game with side-payments where the players are labelled by numbers, $I = \{1, 2, 3\}$, and values of the characteristic function for particular coalitions are

$$v(\emptyset) = 0, \quad v(\{i\}) = 1 \ \text{ for } \ i = 1,2,3, \quad v(\{1,2\}) = 2,$$
$$v(\{1,3\}) = 3, \quad v(\{2,3\}) = 4, \quad v(I) = 6.$$

This game is superadditive and convex, and its core is non-empty, i.e.

$$C = \Big\{ \boldsymbol{x} \in R^3 : x_1 + x_2 + x_3 = 6, \ x_1 \geq 1, \ x_2 \geq 1, \ x_3 \geq 1,$$
$$x_1 + x_2 \geq 2, \ x_1 + x_3 \geq 3, \ x_2 + x_3 \geq 4 \Big\},$$

and, for example, $\boldsymbol{x} = (1, 2, 3) \in C$.

All coalitions $K \subset I$ are therefore effective.

□

The non-emptiness of a core is not strictly connected with superadditivity. There exist superadditive games with an empty core and vice-versa as shown in the following two examples.

Example 3.2 Let us consider a 3-players coalition game with side-payments (I, v), where $I = \{1,2,3\}$, $v(\emptyset) = v(\{i\}) = 0$, for $i = 1,2,3$, $v(\{i,j\}) = 2$, for $i,j \in I$, $i \neq j$, and $v(I) = 2$. This game is superadditive but its core is empty.

□

Example 3.3 Let us consider another 3-players game (I, v), such that $I = \{1,2,3\}$, $v(\emptyset) = 0$, $v(\{1\}) = 0$, $v(\{2\}) = v(\{3\}) = 2$, $v(\{1,2\}) = v(\{1,3\}) = v(\{2,3\}) = 2$, $v(I) = 4$. Its core C is non-empty even if the game is neither superadditive nor convex. For example, vectors $(0,2,2)$ or $(1,2,1)$ or $(2,2,0)$ or $(2,0,2)$ belong to C.

□

An example of a coalition game whose core is not realized by the coalition of all players can be described as follows.

Example 3.4 Let (I, v), where $I = \{1,2,3\}$, be a coalition game with the characteristic function

$$v(\emptyset) = 0, \quad v(\{i\}) = 1 \text{ for } i = 1,2,3, \quad v(\{1,2\}) = 2,$$
$$v(\{1,3\}) = 2, \quad v(\{2,3\}) = 3, \quad v(I) = 3.$$

This game is not superadditive as $v(\{1\}) + v(\{2,3\}) > v(I)$ but its core is non-empty, eual to

$$C = \left\{x \in R^3 : x_1 = 1, \; x_2 + x_3 = 3, \; x_2 \geq 1, \; x_3 \geq 1\right\}$$

and it is realized by the coalition structure $\mathcal{K} = (\{1\}, \{2,3\})$.

□

The core describes a set of possible agreements. That is, agreements which no group of players can effectively object. Generally, a core (if non-empty) can include more imputations, and it is desirable to choose one of them as the best justified. Meanwhile the core (i.e. a possible cooperation) can be derived from basic elements of the game, and the choice of a one-element solution has to be supported by an external concept of "justice". The most widely used one is the vector of Shapley values.

For every player $i \in I$ we denote the Shapley value of game (I, v) for player i by $t_i(v)$. To represent a "just" distribution of the utility among players, the Shapley value is to respect the following three conditions (see [49, 19]):

- The value $t_i(v)$ does not depend on the ordering of players, i.e., on the position of player i in the set I.

- Vector of the Shapley values $(t_i(v))_{i\in I}$ is to be accessible for the coalition of all players I, i.e.

$$\sum_{i\in I} t_i(v) = v(I).$$

- If the set of players I plays two games $(I, v_1), (I, v_2)$, parallelly, with an additive effect, i.e., they form a new game $(I, v_1 + v_2)$ with $(v_1 + v_2)(K) = v_1(K) + v_2(K)$ for any $K \subset I$, then the Shapley values preserve the additivity, i.e.

$$t_i(v_1 + v_2) = t_i(v_1) + t_i(v_2), \quad i \in I.$$

It was shown that such vector of Shapley values is determined by an effective formula.

For every player $i \in I$ his or her *Shapley value* is the number

$$t_i = \sum_{K\subset I} \frac{(n-k)!\,(k-1)!}{n!} \left(v(K) - v(K - \{i\})\right), \tag{3.10}$$

where k is the number of players in K and n is the number of players in I (cf., [46, 49, 19]).

Example 3.5 Let us compute the Shapley values of two of coalition games mentioned in the previous examples.

If (I, v) is the 3-players game in Example 3.1, i.e. $v(\emptyset) = 0$, $v(\{i\}) = 1$, for $i = 1, 2, 3$, $v(\{1, 2\}) = 2$, $v(\{1, 3\}) = 3$, $v(\{2, 3\}) = 4$, $v(I) = 6$, then it is not difficult to verify that

$$t_1 = 3/2, \quad t_2 = 2, \quad t_3 = 5/2.$$

If (I, v) is the 3-players game from Example 3.3, i.e. $v(\emptyset) = 0,\ v(\{1\}) = 0,\ v(\{2\}) = v(\{3\}) = 2,\ v(\{1, 2\}) = v(\{1, 3\}) = v(\{2, 3\}) = 2,\ v(I) = 4$, then the vector of Shapley values is

$$t_1 = 2/3, \quad t_2 = 5/3, \quad t_3 = 5/3.$$

In both the games mentioned the vectors of Shapley values belong to the core of the relevant game. It is also easy to compute the Shapley values for the 3-players game in Example 3.2, i.e. $v(\emptyset) = v(\{i\}) = 0$, for $i = 1, 2, 3$, $v(\{1, 2\}) = v(\{1, 3\}) = v(\{2, 3\}) = v(I) = 2$. The Shapley values form a vector

$(2/3, 2/3, 2/3)$ which is achievable by the all-players coalition I (the sum of its components is 2) but it, evidently, does not fulfill the properties of a core.

We can see that the symmetric possibilities of players 2 and 3 imply the equality of their Shapley values.

□

The previous game, even if its core was empty, is superadditive. The condition of achievability of the Shapley value for coalition I is also valid even if the game is not superadditive and the (nonempty) core is attained by a more complicated coalition structure.

Let us consider a game (I, v) with $I = \{1, 2, 3\}$, $v(\emptyset) = 0$, $v(K) = 2$, for any $K \subset I$, and $K \neq \emptyset$. This game is subadditive, its core is non-empty (it contains exactly one imputation $(2, 2, 2)$) which is attained by the coalition structure of single-players coalitions $\mathcal{K} = (\{1\}, \{2\}, \{3\})$. The vector of the Shapley values is

$$(2/3, 2/3, 2/3).$$

This means that the second condition is fulfilled here, too.

□

The theory of games with side-payments is based on a hidden assumption that there exists a universal representation of utility which can be used for a practical re-distribution of the coalition income among its members. The utility of such a universal representation has to be linear and equal for all players in the given game. This demand makes the assumption about its existence very strong and many times hardly possible to meet. Even money, which is the main "candidate" for this position, does not universally guarantee its fulfilment.

3.2 Coalition Games Without Side–Payments

The next type of coalition games is more general than its name indicates. They, of course, include games without side-payments in which coalition members coordinate their strategies but each of them earns his or her individual pay-off. They also include games with side-payments in which no universal representation of utility exists and in which, consequently, the utility values change during the transmission between players. In this sense, also the previous games with side-payments become a special case of these general games.

Also in this case we denote by I a non-empty and finite set of *players*. By $\mathcal{V}$ we denote a mapping from 2^I into the class of subsets of R^I such that for every *coalition*, there is $K \subset I$ such that

$$\mathcal{V}(K) \text{ is closed },\tag{3.11}$$

$$\text{for } \boldsymbol{x} = (x_i)_{i\in I} \in \mathcal{V}(K),\ \boldsymbol{y} = (y_i)_{i\in I} \in R^I \text{ and } x_i \geq y_i \text{ for all } i \in K \text{ then } \boldsymbol{y} \in \mathcal{V}(K),\tag{3.12}$$

$$\mathcal{V}(K) \neq \emptyset,\tag{3.13}$$

$$\mathcal{V}(K) = R^I \text{ iff } K = \emptyset.\tag{3.14}$$

The mapping $\mathcal{V}$ is called a *general characteristics function.*

If $\boldsymbol{x} = (x_i)_{i\in I} \in R^I$ and $\boldsymbol{y} = (y_i)_{i\in I} \in R^I$ are two *imputations* , then we say that $\boldsymbol{x}$ *dominates* $\boldsymbol{y}$ *via* K coalition $K \subset I$ and write $\boldsymbol{x}\,\mathrm{dom}_K\,\boldsymbol{y}$ iff

$$x_i \geq y_i \text{ for all } i \in K \quad \text{and} \quad x_j > y_j \text{ for some } j \in K.\tag{3.15}$$

For every coalition K we construct a set $\mathcal{V}^*(K) \subset R^I$ such that

$$\mathcal{V}^*(K) = \{\boldsymbol{y} \in R^I : \text{there does not exist } \boldsymbol{x} \in \mathcal{V}(K) \text{ such that } \boldsymbol{x}\,\mathrm{dom}_K\,\boldsymbol{y}\} = \{\boldsymbol{y} \in R^I : \text{ for every } \boldsymbol{x} \in \mathcal{V}(K) \text{ there exists } i \in K \text{ such that } y_i > x_i \text{ or for all } j \in K\ x_j = y_j\}.\tag{3.16}$$

This set $\mathcal{V}^*(K)$ is called the *superoptimum* of coalition K and the intersection

$$P(K) = \mathcal{V}(K) \cap \mathcal{V}^*(K)\tag{3.17}$$

is the *Pareto optimum* of coalition K.

The Pareto optimum of any non-empty coalition need not be non-empty but in the usual non-degenerated cases its non-emptiness is fulfilled. For every coalition $K \subset I$ the set $\mathcal{V}(K)$ is the set of imputations which are accessible for the players in K and $\mathcal{V}^*(K)$ is the set of imputations which cannot be improved (in the sense of domination (3.15)) by any vector of profits accessible for K (cf. [21, 48]).

A *coalition game without side payments* or a *general coalition game* is the pair $(I, \mathcal{V})$. It can be easily seen that the game with side-payments (I, v) presented in Section 3.1 can be described also as a general coalition game $(I, \mathcal{V})$ in which for every non-empty coalition $K \subset I$, there holds

$$\mathcal{V}(K) = \left\{\boldsymbol{x} = (x_i)_{i\in I} : \sum\nolimits_K x_i \leq v(K)\right\}.\tag{3.18}$$

As the sets $\mathcal{V}(K)$ in more general cases are much more flexible than the simple ones fulfilling (3.18), the counterparts of some concepts introduced for the games with side-payments are rather more complex.

We say that a general coalition game $(I, \mathcal{V})$ is *superadditive* iff for any pair of disjoint coalitions $K, L \subset I$, $K \cap L = \emptyset$, the inclusion

$$\mathcal{V}(K \cup L) \supset \mathcal{V}(K) \cap \mathcal{V}(L) \tag{3.19}$$

holds.

Moreover, a game $(I, \mathcal{V})$ is *subadditive* if for any disjoint $K, L \subset I$, we have

$$\mathcal{V}^*(K \cup L) \supset \mathcal{V}^*(K) \cap \mathcal{V}^*(L), \tag{3.20}$$

and it is *additive* if it is both superadditive and subadditive.

A modification of the convexity concept (3.7) for general coalition games is much more complicated (cf. [23]) than the one of superadditivity and subadditivity.

We say that a game $(I, \mathcal{V})$ is *convex* iff for every pair of coalitions $K, L \subset I$ and every pair of imputations

$$\begin{aligned} \boldsymbol{x} &= (x_i)_{i \in I} \in \mathcal{V}(K) \cap \mathcal{V}^*(K \cap L), \\ \boldsymbol{y} &= (y_i)_{i \in I} \in \mathcal{V}(L) \cap \mathcal{V}^*(K \cap L) \end{aligned}$$

any imputation $\boldsymbol{z} \in R^I$ such that

$$z_i = x_i \quad \text{for } i \in K,$$

$$z_i = y_i \quad \text{for } i \in L - K$$

belongs to the set $\mathcal{V}(K \cup L)$.

Evidently, every convex game is superadditive and for any pair of coalitions $K, L \subset I$ in a convex game, we have

$$\mathcal{V}(K) \cap \mathcal{V}(L) \cap \mathcal{V}^*(K \cap L) \subset V(K \cup L).$$

We say that an imputation $\boldsymbol{x} = (x_i)_{i \in I}$ belongs to a *core* $\boldsymbol{C}$ of the game $(I, \mathcal{V})$ iff there exists a coalition structure $\mathcal{K}$ such that $\boldsymbol{x} \in \mathcal{V}(K)$, for all coalitions $K \in \mathcal{K}$ and, moreover, $\boldsymbol{x} \in \mathcal{V}^*(L)$ for all $L \subset I$.

If a game $(I, \mathcal{V})$ is superadditive, then

$$\boldsymbol{C} = \mathcal{V}(I) \cap \left(\bigcap\nolimits_{L \subset I} \mathcal{V}^*(L)\right). \tag{3.21}$$

A formal illustrative examples of coalition games without side-payments are usually analytically rather complicated and, as a consequence, not very easy to follow if the number of players is not quite small. For clarity and visibility, we use in most of the following examples only simple games with a few, cooperating or non-cooperating, players. A 3-players coalition game without side-payments can be described, e.g., as follows.

Example 3.6 Let us consider a 3-players coalition game without side-payment $(I, \mathcal{V})$, where $I = \{1, 2, 3\}$, $\mathcal{V}(\emptyset) = R^3$, and

$$\begin{aligned}
\mathcal{V}(\{i\}) &= \{x \in R^3 : x_i \leq 1/2\}, \quad i = 1,2,3, \\
\mathcal{V}(\{i,j\}) &= \{x \in R^3 : x_i + x_j \leq 1\} \\
&\quad \cap \{x \in R^3 : x_i \leq 1\} \cap \{x \in \mathbb{R}^3 : x_j \leq 1\}, \\
&\qquad i,j = 1,2,3, \ i > j \\
\mathcal{V}(I) &= \{x \in R^3 : x_1 + x_2 + x_3 \leq 2\} \cap \{x \in R^3 : x_1 + x_2 \leq 2\} \cap \\
&\qquad \cap \{x \in \mathbb{R}^3 : x_2 + x_3 \leq 2\} \cap \{x \in \mathbb{R}^3 : x_1 \leq 2\} \cap \\
&\qquad \cap \{x \in \mathbb{R}^3 : x_2 \leq 2\} \cap \{x \in \mathbb{R}^3 : x_3 \leq 2\}.
\end{aligned}$$

Then, it can be seen that the superoptima of coalitions are $V^*(\emptyset) = R^3$,

$$\begin{aligned}
\mathcal{V}^*(\{i\}) &= \{x \in R^3 : x_i \geq 1/2\}, \quad i = 1,2,3, \\
\mathcal{V}^*(\{i,j\}) &= (R^3 - \mathcal{V}(\{i,j\})) \cup \left(\{x \in R^3 : x_i + x_j = 1\} \cap V(\{i,j\})\right), \\
&\qquad \text{for } i = 1,2,3, \ i > j, \\
\mathcal{V}^*(I) &= (R^3 - \mathcal{V}(I)) \cup \left(\{x \in R^3 : x_1 + x_2 + x_3 = 2\} \cap V(I)\right)
\end{aligned}$$

and the Pareto optima $\mathcal{P}(K)$ are

$$\begin{aligned}
\mathcal{P}(\emptyset) &= R^3, \\
\mathcal{P}(\{i\}) &= \{x \in R^3 : x_i = 1/2\}, \quad i = 1,2,3, \\
\mathcal{P}(\{i,j\}) &= \{x \in R^3 : x_i + x_j = 1\} \cap \{x \in R^3 : x_i \leq 1\} \cap \{x \in R^3 : x_j \leq 1\}, \\
&\qquad \text{for } i,j = 1,2,3, \ i > j, \\
\mathcal{P}(I) &= \{x \in R^3 : x_1 + x_2 + x_3 = 2\} \cap \{x \in R^3 : x_1 + x_2 \leq 2\} \cap \\
&\qquad \cap \{x \in R^3 : x_1 + x_3 \leq 2\} \cap \{x \in R^3 : x_2 + x_3 \leq 2\}.
\end{aligned}$$

Then, evidently, $\mathcal{P}(I)$ is the triangular segment of the set $\{x \in R^3 : x_1 + x_2 + x_3 = 2\}$ with vertices $(0,0,2)$ and $(0,2,0)$.

Including further superoptima, we can see that

$$\mathcal{V}(I) \cap \mathcal{V}^*(I) \cap \mathcal{V}^*(\{1,2\}) \cap \mathcal{V}^*(\{1,3\}) \cap \mathcal{V}^*(\{2,3\})$$

is the triangular segment of $\mathcal{P}(I)$ with vertices $(1,1,0)$, $(1,0,1)$ and $(0,1,1)$, and continuing this procedure, we obtain that

$$\mathcal{V}(I) \cap \mathcal{V}^*(I) \cap \mathcal{V}^*(\{1,2\}) \cap \mathcal{V}^*(\{1,3\}) \cap \mathcal{V}^*(\{2,3\}) \cap \mathcal{V}^*(\{1\}) \cap \mathcal{V}^*(\{2\}) \cap \mathcal{V}^*(\{3\})$$

is the triangular subset of $\mathcal{P}(I)$ with vertices $(1,1/2,1/2)$, $(1/2,1,1/2)$ and $(1/2,1/2,1)$.

The final triangular segment is also a core of the considered game because of its superadditivity.

□

We say that for $K \subset I$ the set $\mathcal{V}(K)$ is *sufficient* iff for any $\boldsymbol{x} \in \mathcal{V}(K) - \mathcal{V}^*(K)$ and $\boldsymbol{x} \in \mathcal{V}^*(\{i\})$, for all $i \in I$, there exists $\boldsymbol{y} \in P(K)$ such that $\boldsymbol{y} \operatorname{dom}_K \boldsymbol{x}$ (see [23]).

Even in games without side-payments the convexity is a sufficient condition for the non-emptiness of the core under the assumption that all sets $\mathcal{V}(K)$, $K \subset I$, are sufficient as shown in [23].

We say that a game $(I, \mathcal{V})$ is *balanced* if for every balanced set of coalitions $\mathcal{L} \subset 2^I$ (cf. (3.8)) the inclusion

$$\bigcap\nolimits_{L \in \mathcal{L}} \mathcal{V}(L) \subset \mathcal{V}(I) \tag{3.22}$$

holds.

Even in this case the balancedness of a game $(I, \mathcal{V})$ implies its superadditivity and it is a sufficient (in this case not necessary) condition for the non-emptiness of the core (3.21) under the assumption that, for every coalition structure $\mathcal{K}$, there is

$$\left(\bigcap\nolimits_{i \in I} \mathcal{V}^*(\{i\})\right) \cap \left(\bigcap\nolimits_{K \in \mathcal{K}} \mathcal{V}(K)\right).$$

The general form of games without side-payments admits a relatively wide variability of their particular cases. Let us illustrate some of them on examples of 2-players games.

Example 3.7 Let $I = \{1, 2\}$ be a set of players who can proceed either individually or in a cooperative way. If they proceed individually, then Player 1 winns 2 units of utility and Player 2 winns 1 unit. If they cooperate their behaviour, then they can use one of three accessible coalitional strategies, and they win the following units of utility:

strategy of I	Player 1 wins	Player 2 wins
s_1	1	4
s_2	3	3
s_3	4	2

There does not exist a possibility to mix or combine the coalitional strategies or to transfer the utilities. Then, formally, we have

$$\begin{aligned}
\mathcal{V}(\{1\}) &= \{\boldsymbol{x} \in R^2 : x_1 \leq 2\}, \quad \mathcal{V}(\{2\}) = \{\boldsymbol{x} \in R^2 : x_2 \leq 1\}, \\
V(I) &= \{\boldsymbol{x} \in R^2 : x_1 \leq 1 \text{ and } x_2 \leq 4\} \cup \{\boldsymbol{x} \in R^2 : x_1 \leq 3,\ x_2 \leq 3\} \cup \\
&\quad \cup \{\boldsymbol{x} \in R^2 : x_1 \leq 4,\ x_2 \leq 2\}.
\end{aligned}$$

This game is superadditive, i.e.

$$\begin{aligned}
\mathcal{V}^*(\{1\}) &= \{\boldsymbol{x} \in R^2 : x_1 \geq 2\}, \quad \mathcal{V}^*(\{2\}) = \{\boldsymbol{x} \in R^2 : x_2 \geq 1\}, \\
\mathcal{V}^*(I) &= (R^2 - \mathcal{V}(I)) \cup (\{(1,3), (3,3), (4,2)\})
\end{aligned}$$

and the Pareto optima are

$$\begin{aligned}\mathcal{P}(\{1\}) &= \{\boldsymbol{x} \in R^2 : x_1 = 2\}, \quad \mathcal{P}(\{2\}) = \{\boldsymbol{x} \in R^2 : x_2 = 1\}, \\ \mathcal{P}(I) &= \{(1,3),(3,3),(4,2)\}.\end{aligned}$$

It means that the game is not subadditive and its core is a subset of $\mathcal{P}(I)$, namely

$$\boldsymbol{C} = \{(3,3),(4,2)\}.$$

□

Example 3.8 Let us consider the game described in the preceeding Example 3.7 but let us assume a probabilistic mixtures of the coalitional strategies, which we denote by

$$\lambda_1 s_2 + \lambda_2 s_2 + \lambda_3 s_3, \quad \lambda_i \geq 0, \quad i = 1,2,3, \quad \lambda_1 + \lambda_2 + \lambda_3 = 1,$$

and let the utilities attained by players in I be in such a case

$$\begin{array}{ll} 1 \cdot \lambda_1 + 3 \cdot \lambda_2 + 4 \cdot \lambda_3 & \text{for Player 1} \\ 3 \cdot \lambda_1 + 3 \cdot \lambda_2 + 2 \cdot \lambda_3 & \text{for Player 2.} \end{array}$$

Then, evidently, the sets $\mathcal{V}(\{i\})$, $\mathcal{V}^*(\{i\})$ and $\mathcal{P}(\{i\})$, for $i = 1,2$, are identical with those in Example 3.7.

But

$$\begin{aligned}\mathcal{V}(I) &= \{\boldsymbol{x} \in R^2 : x_2 \leq 4\} \cap \{\boldsymbol{x} \in R^2 : x_1 \leq 4\} \\ &\qquad \cap \{\boldsymbol{x} \in R^2 : x_1 + x_2 \leq 6\} \cap \{\boldsymbol{x} \in R^2 : x_1 + 2x_2 \leq 9\}, \\ \mathcal{V}^*(I) &= (R^2 - \mathcal{V}(I)) \cup \{\boldsymbol{x} \in R^2 : x_1 + x_2 = 6\} \cup \{\boldsymbol{x} \in R^2 : x_1 + 2x_2 = 9\}, \\ \mathcal{P}(I) &= \{\boldsymbol{x} \in R^2 : x_1 + x_2 \leq 6,\ 3 \leq x_1 \leq 4\} \\ &\qquad \cup \{\boldsymbol{x} \in R^2 : x_1 + 2x_2 = 9,\ 1 \leq x_1 \leq 3\}.\end{aligned}$$

This game is evidently superadditive but not subadditive and its core $\boldsymbol{C} \subset \mathcal{P}(I)$ is equal to

$$\{\boldsymbol{x} \in R^2 : x_1 + x_2 = 6,\ 3 \leq x_1 \leq 4\} \cup \{\boldsymbol{x} \in R^2 : x_1 + 2x_2 = 9,\ 2 \leq x_1 \leq 3\}.$$

□

Relations between the superadditivity, subadditivity and additivity can be illustrated by the following two examples.

Example 3.9 Let $(I, \mathcal{V})$ be a 2-players coalition game where $I = \{1, 2\}$ and

$$\begin{aligned}\mathcal{V}(I) &= \{\boldsymbol{x} \in R^2 : x_1^2 + x_2^2 \leq 1,\ 0 \leq x_1 \leq 1,\ 0 \leq x_2 \leq 1\} \cap \\ &\cap \{\boldsymbol{x} \in R^2 x_1 \leq 1\} \cap \{\boldsymbol{x} \in R^2 x_2 \leq 1\}.\end{aligned}$$

Then,

$$\mathcal{V}^*(I) = (R^2 - \mathcal{V}(I)) \cup \{\boldsymbol{x} \in R^2 : x_1 + x_2 = 1,\ 0 \leq x_1,\ 0 \leq x_2\}$$

and

$$\mathcal{P}(I) = \mathcal{V}(I) \cap \mathcal{V}^*(I) = \{\boldsymbol{x} \in R^2 : x_1^2 + x_2^2 = 1,\ 0 \leq x_1,\ 0 \leq x_2\}.$$

Let us consider two real numbers v_1, v_2 and let, for both $i = 1, 2$,

$$\mathcal{V}(\{i\}) = \{\boldsymbol{x} \in R^2 : x_i \leq v_i\}$$

which implies

$$\mathcal{V}^*(\{i\}) = \{\boldsymbol{x} \in R^2 : x_i \geq v_i\}.$$

It is evident that $(I, \mathcal{V})$ is superadditive iff $(v_1, v_2) \in \mathcal{V}(I)$, it is subadditive iff $(v_1, v_2) \in \mathcal{V}^*(I)$, and it is additive iff $(v_1, v_2) \in \mathcal{P}(I)$.

□

Example 3.10 Let $(I, \mathcal{V})$ be 2-players coalition game such that $I = \{1, 2\}$ and

$$\mathcal{V}(I) : \{\boldsymbol{x} \in R^2 : x_1 \cdot x_2 \leq 1,\ x_1 > 0\} \cup \{\boldsymbol{x} \in R^2 : x_1 \leq 0\}.$$

Then

$$\begin{aligned}\mathcal{V}^*(I) &= \{\boldsymbol{x} \in R^2 : x_1 \cdot x_2 \geq 1,\ x_1 > 0\} \\ \mathcal{P}(I) &= \mathcal{V}(I) \cap \mathcal{V}^*(I) = \{\boldsymbol{x} \in R^2 : x_1 \cdot x_2 = 1,\ x_1 > 0\}.\end{aligned}$$

Let, analogously to the previous example, $v_1, v_2 \in R$ be real numbers for which

$$\mathcal{V}(\{i\}) = \{\boldsymbol{x} \in R^2 : x_i \leq v_i\}, \quad \mathcal{V}^*(\{i\}) = \{\boldsymbol{x} \in R^2 : x_i \geq v_i\}, \quad i = 1, 2.$$

Then, $(I, \mathcal{V})$ is superadditive iff $(v_1, v_2) \in \mathcal{V}(I)$, it is subadditive iff $(v_1, v_2) \in \mathcal{V}^*(I)$, and it is additive iff $(v_1, v_2) \in \mathcal{P}(I)$.

□

The coalition games with side-payments can also be considered as a special case of the game without side-payment with lineary bounded sets $\mathcal{V}(I)$ (see also (3.18)).

Example 3.11 Let I be a set of players and let for each coalition $K \subset I$ exist a real number $v(K) \in R$. Then the sets $\mathcal{V}(K),\ K \subset I$, defined by

$$\mathcal{V}(K) = \{\boldsymbol{x} \in R^I : \sum_K x_i \leq v(K)\}$$

fulfil properties of the general characteristic function and, consequently, $(I, \mathcal{V})$ is a coalition game without side payments with

$$\mathcal{V}^*(K) = \{\boldsymbol{x} \in R^I : \sum_K x_i \geq v(K)\}, \quad \mathcal{P}(K) = \{\boldsymbol{x} \in R^I : \sum_K x_i = v(K)\}$$

for each $K \subset I$.

□

Any subadditive game as, e.g., following the relevant cases of v_1, v_2 in Examples 3.9 and 3.10, illustrates the cases of game where the core is non-empty, realized by a coalition structure $(\{1\}, \{2\})$ and not necessarily realized by the coalition of all players I (if the game is not additive). An example of a 3-players game where the core is realized by another coalition structure than the one formed by single all-players coalition is given below.

Example 3.12 Let us consider a 3-players coalition game $(I, \mathcal{V})$ with $I = \{1, 2, 3\}$, $\mathcal{V}(\emptyset) = R^3$ and

$$\begin{aligned}
\mathcal{V}(\{1\}) &= \{\boldsymbol{x} \in R^3 : x_1 \leq \sqrt{2}/2\}, \\
\mathcal{V}(\{2\}) &= \{\boldsymbol{x} \in R^3 : x_2 \leq 1/2\}, \quad \mathcal{V}(\{3\}) = \{\boldsymbol{x} \in R^3 : x_3 \leq 1/2\}, \\
\mathcal{V}(\{1,2\}) &= \{\boldsymbol{x} \in R^3 : x_1 + x_2 \leq \sqrt{2}\} \\
&\quad \cap \{\boldsymbol{x} \in R^3 : x_1 \leq 1\} \cap \{\boldsymbol{x} \in R^3 : x_2 \leq 1\}, \\
\mathcal{V}(\{1,3\}) &= \{\boldsymbol{x} \in R^3 : x_1 + x_3 \leq \sqrt{2}\} \\
&\quad \cap \{\boldsymbol{x} \in R^3 : x_1 \leq 1\} \cap \{\boldsymbol{x} \in R^3 : x_3 \leq 1\}, \\
\mathcal{V}(\{2,3\}) &= \{\boldsymbol{x} \in R^3 : x_2 + x_3 \leq \sqrt{2}\} \\
&\quad \cap \{\boldsymbol{x} \in R^3 : x_2 \leq 1\} \cap \{\boldsymbol{x} \in R^3 : x_3 \leq 1\}.
\end{aligned}$$

If we denote

$$\mathcal{V}_1 = \{\boldsymbol{x} \in R^3 : x_1^2 + x_2^2 + x_3^2 \leq 1,\ x_1 \geq 0,\ x_2 \geq 0,\ x_3 \geq 0\}$$

then

$$\mathcal{V}(I) = \mathcal{V}_1 \cup \{\boldsymbol{x} \in R^3 : \exists \boldsymbol{y} \in V_1,\ y_1 \geq x_1,\ y_2 \geq x_2,\ y_3 \geq x_3\}.$$

This game is not superadditive as, e.g., $\mathcal{V}(\{1\}) \cap \mathcal{V}(\{2,3\})$ is not a subset of $\mathcal{V}(I)$ as the imputation $(\sqrt{2}/2,\ \sqrt{2}/2,\ \sqrt{2}/2)$ (and some others) belongs to the intersection mentioned but not to $\mathcal{V}(I)$.

The core of this game is attained by the coalition structure $(\{1\}, \{2,3\})$ and it is the set

$$\mathcal{V}(\{1\}) \cap \mathcal{V}(\{2,3\}) \cap \mathcal{V}^*(\{1\}) \cap \mathcal{V}^*(\{2,3\}) \cap \mathcal{V}^*(\{1,2\}) \cap \mathcal{V}^*(\{1,3\}) =$$
$$= \left\{(\sqrt{2}/2, \sqrt{2}/2, \sqrt{2}/2)\right\}.$$

□

The concept of effectiveness of coalitions, mentioned in the games with side-payments, can be considered even in this case. Also in the games without side-payments any coalition $K \subset I$ is said to be effective iff there exists an imputation $\boldsymbol{x}_K \in R^K$ which is accessible for K, i.e. $\boldsymbol{x}_K \in \mathcal{V}(K)$, and which cannot be protested by any subcoalition $J \subset K$, i.e. $\boldsymbol{x}_K \in \mathcal{V}^*(J)$, for all $J \subset K$. It is easy to see that I is effective if and only if the core C given by (3.21) is non-empty.

4. Vagueness and Its Processing

After having introduced basic concepts of the deterministic coalition game theory we can discuss possible sources of vagueness entering the model.

It is generally acceptable to assume that real cooperative situations are not as deterministic as the models presented above indicate. The book [15] and some related works (e.g., [11, 12, 13, 22])study situations in which a player can simultaneously participate in more coalitions with different "intensities" of participation. Formulated in terms of fuzzy set theory, coalitions become fuzzy subsets of the set I where an "intensity" of participation determines the membership functions. The results obtained in the framework of this theory are interesting and inspiring. It could be noted that a model of parallel realization of several coalition structures with different weights (but not in terms of fuzzy set theory) was already mentioned in [22].

In the following sections we will develop another type of vagueness in coalition games. It regards the motivation of players to form coalitions. The motivation is given by an expectation of individual profits, or – in the language of the previous sections – by values of the characteristic functions. As we consider two types of coalitions games – those with and without side-payments – differing exactly in the formal model of the characteristic function, we also need to develop two approaches to their fuzzification.

A characteristic function v in the games with side-payments maps the set of coalitions into the set of real numbers. The values $v(K)$ for $K \subset I$ represent the expected total income of coalitions and the important methods and results of the theory regard manipulations with these numbers. If the players' and coalitions' expectations of profits are vague, then they can be modelled by fuzzy quantities and operations on them are the main tools for their processing. If we wish to derive analogies of the concepts and results briefly summarized in Section 3.1, we have to process these fuzzy quantities according to the methods briefly summarized in Section 2 and in more details summarized, e.g., in [25]. The formalism presented in Section 3.1 shows that the most frequent operations on fuzzy quantities used in the analysis of the games with fuzzy characteristic function will be the addition and comparison of values. A practical application of these methods can indicate a serious limitation of their use. Namely, each algebraical operation with fuzzy quantities which is based on the extension principle increases uncertainty of the results. An alternative approach to fuzzy quantities avoiding this problem can be based on methods explained in [38] and investigated in several other papers. This approach is briefly summarized in Section 16.

On the other hand, values of the general characteristic function $\mathcal{V}$ in the coalition games without side-payments are subsets of the multidimensional real space R^I. It means that the sets of vaguely expected imputations will be represented by fuzzy subsets of R^I. The most frequent operations used during their processing will be the intersection and inclusion of fuzzy sets. The elementary operations on fuzzy sets seem to be simpler than thoseon fuzzy numbers and, consequently, the fuzzification of the general coalition games may look easier than the one of the games with side-payments.

We will see that there exist in this respect some methodological difficulties, especially with the domination relation and its transmission to the elements of fuzzy sets. These difficulties are based on the fact that the processing of the coalition games without side-payments and vague profits combines the processing of fuzzy sets in the set theoretical sense and imputations as crisp multidimensional vectors. And an alternative approach to the domination relation in fuzzy coalition games is briefly discussed in Section 14.

Part II

FUZZY BARGAINING WITH SIDE–PAYMENTS

In this part we describe and discuss the coalition games with side-payments in which the expected incomes of coalitions are vague. This vagueness of profits implies also a vague validity of other properties and characteristics of the game and, moreover, a vague belongingness of some imputations to the core. Heuristically, we presume that the original, most expected, values of the characteristic function are extended by other possible values.

Formally, let us consider a (deterministic) coalition game with side-payments (I, v) and let $w(K)$ for any $K \subset I$ be a fuzzy quantity, $w(K) \in I\!R$, with the membership function $\mu_K : R \to [0, 1]$ such that:

— $\mu_K(v(K)) = 1$,

— $\mu_K(x)$ is nondecreasing for $x < v(K)$ and nonincreasing for $x > v(K)$, and

— $\mu_\emptyset(x) = 1$ for $x = 0$ and $\mu_\emptyset(x) = 0$ for $x \neq 0$, i.e. $w(\emptyset) = \langle 0 \rangle$.

Then we call the mapping $w : 2^I \to I\!R$ a *fuzzy characteristic function* and the pair (I, w) a *fuzzy extension of the coalition game* (I, v).

These fuzzy extensions of the coalition games represent the type of *fuzzy coalition games* in which we are interested in this part.

A fuzzy intension of a coalition game with side-payments can be illustrated, e.g., on the game introduced in Example 3.3, page 21.

Example 4.1 Let (I, v) be a 3-players coalition game with side-payments with $I = \{1, 2, 3\}$, $v(\emptyset) = 0$, $v(\{1\}) = 0$, $v(\{2\}) = v(\{3\}) = 2$, $v(\{1, 2\}) = v(\{1, 3\}) = v(\{2, 3\}) = 2$, and $v(I) = 4$.

One of its possible fuzzy extensions is (I, w) where

$$\begin{array}{lll} \text{for } w(\emptyset), & \mu_\emptyset(0) = 1, & \mu_\emptyset(x) = 0 \quad \text{for } x \neq 0, \\ \text{for } w(\{1\}), & \mu_{\{1\}}(x) = 0 & \text{for } x < 0 \text{ or } x > 1 \\ & \mu_{\{1\}}(x) = 1 - x & \text{for } x \in [0, 1], \\ \text{for } w(\{2\}), & \mu_{\{2\}}(x) = 1 & \text{for } x = 2, \quad \mu_2(x) = 0 \quad \text{for } x \neq 2 \end{array}$$

(it means that the pay-off of the 1-player coalition $\{2\}$ is exactly known), and

$$\begin{array}{lll} \text{for } w(\{3\}), & \mu_{\{3\}}(x) = x \quad 1 & \text{for } x \in [1, 2], \\ & \mu_{\{3\}}(x) = 3 - x & \text{for } x \in [2, 3], \\ & \mu_{\{3\}}(x) = 0 & \text{for } x < 1 \text{ or } x > 3, \\ \text{for } w(\{1, 2\}), & \mu_{\{1,2\}}(x) = x/2 & \text{for } x \in [0, 2], \\ & \mu_{\{1,2\}}(x) = 0 & \text{for } x < 0 \text{ or } x > 2, \\ \text{for } w(\{1, 3\}), & \mu_{\{1,3\}}(x) = x - 1 & \text{for } x \in [1, 2], \\ & \mu_{\{1,3\}}(x) = 2 - x/2 & \text{for } x \in [2, 4], \\ & \mu_{\{1,3\}}(x) = 0 & \text{for } x < 1 \text{ or } x > 4, \end{array}$$

$$
\begin{array}{lll}
\text{for } w(\{2,3\}), & \mu_{\{2,3\}}(x) = 1 & \text{for } x \in (1,2], \\
& \mu_{\{2,3\}}(x) = 1/3 & \text{for } x \in (2,3], \\
\\
\text{for } w(I), & \mu_I(x) = (x-3)^2 & \text{for } x \in [3,4], \\
& \mu_I(x) = \sqrt{1-(x-4)^2} & \text{for } x \in [4,5], \\
& \mu_I(x) = 0 & \text{for } x < 3 \text{ or } x > 5.
\end{array}
$$

□

In this part, the application of relations briefly recalled in Section 2 represents a methodological background for the processing of fuzzy game-theoretical objects.

5. Fuzzy Additivity and Related Topics

Simple concepts of superadditivity and subadditivity belong to the basic ones in the theory of coalition games. They evidently influence the readiness of players to cluster into effectively cooperating coalitions. Results regarding this topic are briefly presented in some of the referred papers as, e.g., in [32], where the most detailed analysis of the problem is introduced, and where the formal proofs of statements that are rather less formally presented below, are given.

Similarities between a formal structure of the definitions of superadditivity, subadditivity, additivity and convexity lead to an essential similarity between results regarding the properties of their fuzzy counterparts introduced in the following subsections of this section.

It is intuitively obvious and possible to to conclude that the properties investigated in this section become fuzzy properties if a deterministic game (I, v) is substituted by its fuzzy extension (I, w). The possibilities of their validity can be derived from the membership functions μ_K, $K \subset I$.

5.1 Fuzzy Superadditivity

The concept of superadditivity in the fuzzy coalition games demands a deeper analysis than the one in the deterministic games. Namely, while the deterministic superadditivity is based on the comparison of scalar values – in terms of fuzzy quantites the modal values of membership functions – its fuzzy counterpart is based on the comparison of whole shapes of membership functions. We will see that this difference essentially influences the correspondence between the intuitive meaning of superadditivity and its formal definition.

To be more specific, as in the deterministic case, the inequality

$$v(K) + v(L) \leq v(K \cup L),$$

for any disjoint pair of coalitions, obviously implies the inequality

$$v(K_1) + v(K_2) + \cdots + v(K_m) \leq v(K_1 \cup K_2 \cup \cdots \cup K_m),$$

for any disjoint set of coalitions $\{K_1, \ldots, K_m\}$, in the fuzzy case it may not be so simple, namely, the possibility of

$$w(K \cup L) \succsim w(K) \oplus w(L)$$

for all $K, L \subset I$, $K \cap L \neq \emptyset$, need not be smaller than the possibility of

$$w(K_1 \cup \cdots \cup K_m) \succsim w(K_1) \oplus \cdots \oplus w(K_m)$$

as shown in the following example.

Example 5.1 Let us consider a 3-players coalition game (I, v) with the characteristic function v, where $I = \{1, 2, 3\}$, and

$$\begin{array}{l} v(\{i\}) = 4 \quad \text{for } i = 1,2,3 \\ v(\{1,2\}) = v(\{1,3\}) = 6, \quad v(\{2,3\}) = 8 \\ v(I) = 11. \end{array}$$

This game is neither superadditive nor subadditive because, on the one hand,

$$v(\{1\}) + v(\{2,3\}) > v(I) \quad \text{and} \quad v(\{1\}) + v(\{2\}) + v(\{3\}) > v(I),$$

and, on the other hand,

$$v(\{3\}) + v(\{1,2\}) < v(I) \quad \text{and} \quad v(\{2\}) + v(\{1,3\}) < v(I).$$

Let us consider its fuzzy extension (I, w), where

$$\begin{array}{rcll} \mu_{\{i\}}(x) & = & 10x - 39 & \text{for} \quad x \in [3.9, 4], \\ & = & 5 - x & \text{for} \quad x \in [4, 5], \\ & = & 0 & \text{otherwise}, \end{array}$$

for $i = 1, 2, 3$, and further

$$\begin{array}{rcll} \mu_{\{1,2\}}(x) = \mu_{\{1,3\}}(x) & = & x - 5 & \text{for} \quad x \in [5, 6], \\ & = & \frac{5}{2} - \frac{1}{4}x & \text{for} \quad x \in [6, 10], \\ & = & 0 & \text{for} \quad x \notin [5, 10], \end{array}$$

$$\begin{array}{rcll} \mu_{\{2,3\}}(x) & = & \frac{1}{2}x - 3 & \text{for} \quad x \in [6, 8], \\ & = & 5 - \frac{1}{2}x & \text{for} \quad x \in [8, 10], \\ & = & 0 & \text{for} \quad x \notin [6, 10], \end{array}$$

$$\begin{array}{rcll} \mu_I(x) & = & x - 10 & \text{for} \quad x \in [10, 11], \\ & = & 12 - x & \text{for} \quad x \in [11, 12], \\ & = & 0 & \text{for} \quad x \notin [10, 12]. \end{array}$$

Let us denote by $\mu_{\{1\},\{2\},\{3\}} : R \to [0, 1]$ the membership function of the sum $w(\{1\}) \oplus w(\{2\}) \oplus w(\{3\})$ and by $\mu_{\{1\},\{2,3\}} : R \to [0, 1]$ the membership function of $w(\{1\}) \oplus w(\{2, 3\})$.

So, we have

$$\begin{aligned} \mu_{\{1\},\{2\},\{3\}}(x) &= \tfrac{10}{3}\,x - 39 && \text{for} \quad x \in [11.7, 12], \\ &= 5 - \tfrac{1}{3}\,x && \text{for} \quad x \in [12, 15], \\ &= 0\text{otherwise}, \end{aligned}$$

$$\begin{aligned} \mu_{\{1\},\{2,3\}}(x) &= \tfrac{10}{21}\,x - \tfrac{99}{21} && \text{for} \quad x \in [9.9, 12], \\ &= 5 - \tfrac{1}{3}\,x && \text{for} \quad x \in [12, 15], \\ &= 0 && \text{for} \quad x \notin [9.9, 15]. \end{aligned}$$

Then, the possibility of

$$w(I) \succsim w(\{1\}) \oplus w(\{2,3\})$$

is

$$\nu_{\succsim}\,(w(I), w(\{i\}) \oplus w(\{2,3\})) = \frac{21}{31},$$

while the possibility of

$$w(I) \succsim w(\{1\}) \oplus w(\{2\}) + w(\{3\})$$

is

$$\nu_{\succsim}\,(w(I), w(\{1\}) \oplus w(\{2\}) \oplus w(\{3\})) = \frac{3}{13}.$$

□

Let us start with a close fuzzy analogy of the deterministic superadditivity which we call a weak one.

Let (I, v) be a coalitional game with side-payments and (I, w) be its fuzzy extension. Then we say that (I, w) is *simply fuzzy superadditive* if for any pair of disjoint coalitions $K, L \subset I$, $K \cap L = \emptyset$, there is

$$w(K \cup L) \succsim w(K) \oplus w(L). \tag{5.1}$$

The ordering relation (5.1) is valid with a possibility $\nu_{\succsim}(w(K \cup L), w(K) \oplus w(L))$ given by (2.9), and briefly denoted by $\overline{\nu}(K, L)$, and

$$\begin{aligned} \nu_{\succsim}(w(K \cup L), w(K) \oplus w(L)) &= \overline{\nu}(K, L) = \\ &= \sup_{\substack{x, y \in R \\ x \geq y}} [\min(\mu_{K \cup L}(x), \mu_{K+L}(y))] \end{aligned} \tag{5.2}$$

where the symbol μ_{K+L} denotes the membership function of the fuzzy quantity $w(K) + w(L)$.

This means, due to (2.3), for $y \in R$, that

$$\mu_{K+L}(y) = \sup_{z\in R} [\min(\mu_K(z), \mu_L(y-z))] . \tag{5.3}$$

Using (5.2) for all pairs of disjoint coalitions we can easily derive the possibility that the fuzzy game (I, w) is simply superadditive as the number

$$\nu^0_{\text{super}}(I, w) = \min (\overline{\nu}(K, L) : K, L \subset I, K \cap L = \emptyset) . \tag{5.4}$$

Evidently, for disjoint coalitions, we have $K, L, \overline{\nu}(K, L) = \overline{\nu}(L, K)$. Moreover, if (I, w) is a fuzzy extension of (I, v), if $K, L \subset I$, $K \cap L = \emptyset$ and if $v(K \cup L) \geq v(K) + v(L)$, then $\overline{\nu}(K, L) = 1$ which follows from (5.2).

This means that for any fuzzy extension (I, w) of a superadditive game (I, v), the equality $\nu^0_{\text{super}}(I, w) = 1$ holds.

The previous implication shows that the simple fuzzy superadditivity in (I, w) is an extension of the deterministic superadditivity concept in (I, v).

Intuitively, we can expect, and formally prove, that that the more uncertainty is included in the fuzzy extension of a game the higher is the possibility of its superadditivity. Namely, if (I, w) and (I, w') are fuzzy extensions of (I, v), if we denote by μ_K and μ'_K the membership functions of $w(K)$ and $w'(K)$, respectively, and if $\mu_K(x) \geq \mu'_K(x)$, for all $K \subset I$ and $x \in R$, and if we, finally, denote by $\overline{\nu}(K, L)$ and $\overline{\nu}'(K, L)$ the values (5.2) for $K, L \subset I$, $K \cap L = \emptyset$, then $\overline{\nu}(K, L) \geq \overline{\nu}'(K, L)$ and $\nu^0_{\text{super}}(I, w) \geq \nu^0_{\text{super}}(I, w')$.

The previous property can be extremely meaningful if the uncertainty connected with (I, w) is significant enough. Namely, for every deterministic game (I, v) there exists its fuzzy extension (I, w) such that for every pair of disjoint coalitions K, L, there is $\overline{\nu}(K, L) = 1$ and, consequently, $\nu^0_{\text{super}}(I, w) = 1$, as well.

This statement can be easily proven (cf. [32]) if we choose the fuzzy quantities $w(K)$ for which the sets

$$\{x \in R : \mu_K(x) = 1\}$$

are large enough.

On the other hand, the result mentioned above is not true if we keep the principle that for every $K \subset I$, $\mu_K(x) = 1$, for $x = v(K)$ only. For any fuzzy extension (I, w) of (I, v) respecting this condition and for any pair $K, L \subset I$, $K \cap L = \emptyset$, $\overline{\nu}(K, L) = 1$ iff $v(K \cup L) \geq v(K) + v(L)$ and, consequently, $\nu^0_{\text{super}}(I, w) = 1$ iff (I, v) is superadditive in the deterministic sense. However, the previous results mean that $\nu^0_{\text{super}}(I, w)$ can be positive even if (I, v) is not superadditive.

The notes formulated in the introductory paragraphs of this section and illustrated by Example refexample20 show that in case of the fuzzy coalition games with side-payments the simple superadditivity does not imply the universal superadditivity regarding the union of an arbitrary class of disjoint coalitions.

We say that an extension (I, w) of a coalition game (I, v) is *fuzzy superadditive* if for any class of disjoint coalitions $\{K_1, \dots, K_m\}$, $K_j \cap K_k = \emptyset$, for $j \neq k$, $j, k = 1, \dots, m$, there holds

$$w(K_1 \cup \cdots \cup K_m) \succsim w(K_1) \oplus \cdots \oplus w(K_m). \tag{5.5}$$

Analogously to the previous two-coalition case, we denote the possibility of (5.5) by

$$\overline{\nu}(K_1, \dots, K_m) = \nu_{\succsim}(w(K_1 \cup \cdots \cup K_m),\, w(K_1) \oplus \cdots \oplus w(K_m))$$

and the possibility that (I, w) is fuzzy superadditive will be denoted by

$$\begin{aligned}\nu_{\text{super}}(I, w) \;=\; & \min\left(\overline{\nu}(K_1, \dots, K_m) : (K_1, \dots, K_m),\ m \geq 2,\ K_j \subset I\right. \\ & \left. K_j \cap K_k = \emptyset,\ \text{for } j \neq k, j, k = 1, \dots, m\right). \end{aligned} \tag{5.6}$$

It is evident that, generally,

$$\nu^0_{\text{super}}(I, w) \geq \nu_{\text{super}}(I, w),$$

and that the properties of $\nu^0_{\text{super}}(I, w)$ given above are valid for $\nu_{\text{super}}(I, w)$, as well. Namely, if (I, v) is superadditive in the deterministic sense and if (I, w) is its fuzzy extension, then $\nu_{\text{super}}(I, w) = 1$. Moreover, if we preserve the condition of $\mu_K(x) = 1$ iff $x = v(K)$, for all $K \subset 2^I$, then $\nu_{\text{super}}(I, w) = 1$ iff (I, v) is superadditive. If we are not limited by this condition, then for each coalition game (I, v) there exists its fuzzy extension (I, w) such that (I, w) is fuzzy superadditive with possibility $\nu_{\text{super}}(I, w) = \nu^0_{\text{super}}(I, w) = 1$.

The difference between the simple fuzzy superadditivity $\nu^0_{\text{super}}(I, w)$ and the fuzzy superadditivity $\nu_{\text{super}}(I, w)$ can be seen even in the game discussed in Example 5.1.

Example 5.2 Let us consider a coalition game (I, v) and its fuzzy extension (I, w) presented in the previous Example 5.1. Let us compute the possibilities $\overline{\nu}(K, L)$ and $\overline{\nu}(\{1\}, \{2\}, \{3\})$, $K, L \subset I$, $K \cap L = \emptyset$. Then it is easy to see that

$$\overline{\nu}(\{1\}, \{2\}) = \frac{11}{21}, \quad \overline{\nu}(\{1\}, \{3\}) = \frac{11}{21}, \quad \overline{\nu}(\{2\}, \{3\}) = 1$$

since, for $i, j \in I$,

$$\begin{aligned}\mu_{w(\{i\}) \oplus w(\{j\})}(x) &= 5x - 39 && \text{for} && x \in [7.8, 8], \\ &= 5 - \tfrac{1}{2}x && \text{for} && x \in [8, 10], \\ &= 0 && \text{otherwise.}\end{aligned}$$

Analogously, we have

$$\overline{\nu}(\{1\}, \{2, 3\}) = \frac{21}{31}, \quad \overline{\nu}(\{2\}, \{1, 3\}) = 1,$$

$$\overline{\nu}(\{3\}, \{1, 2\}) = 1, \quad \overline{\nu}(\{1\}, \{2\}, \{3\}) = \frac{3}{13}.$$

This means that

$$\nu^0_{\text{super}}(I,w) = \min\Big(\overline{\nu}(\{1\},\{2\}),\overline{\nu}(\{1\},\{3\}),\overline{\nu}(\{2\},\{3\}),\overline{\nu}(\{1\},\{2,3\}), \overline{\nu}(\{2\},\{1,3\}),\overline{\nu}(\{3\},\{1,2\})\Big) = \frac{11}{21},$$

while

$$\nu_{\text{super}}(I,w) = \min\left(\nu^0_{\text{super}}(I,w),\overline{\nu}(\{1\},\{2\},\{3\})\right) = \frac{3}{13}.$$

□

The situation in which $\nu_{\text{super}}(I,w) < \nu^0_{\text{super}}(I,w)$, illustrated by the previous example, need not appear in each fuzzy coalition game. The relation between $\overline{\nu}(J,K,L)$ and $\overline{\nu}(J,K\cup L)$, for disjoint triple of coalitions J,K,L, can be either $\overline{\nu}(J,K,L) < \overline{\nu}(J,K\cup L)$ as shown in Example 5.2 but also $\overline{\nu}(J,K,L) = \overline{\nu}(J,K\cup L)$ and $\overline{\nu}(J,K,L) > \overline{\nu}(J,K\cup L)$.

The last of the two possibilities are illustrated in the following examples.

Example 5.3 Let us consider a 3-players coalition game (I,v) with $I = \{1,2,3\}$, $v(\{i\}) = 2$, $v(\{i,j\}) = 4$ for $i,j \in I$, $i \neq j$, and $v(I) = 5$. Let (I,w) be its fuzzy extension, where the membership functions μ_K of $w(K)$, $K \subset I$, are

$$\begin{aligned}
\mu_{\{i\}}(x) &= x-1 && \text{for } x \in [1,2],\ i=1,2,3,\\
&= 3-x && \text{for } x \in [2,3],\\
&= 0 && \text{for } x \notin [1,3],
\end{aligned}$$

$$\begin{aligned}
\mu_{\{i,j\}}(x) &= \tfrac{1}{2}x-1 && \text{for } x \in [2,4],\ i,j \in I,\ i<j,\\
&= 5-x && \text{for } x \in [4,5],\\
&= 0 && \text{for } x \notin [2,5],
\end{aligned}$$

$$\begin{aligned}
\mu_I(x) &= x-4 && \text{for } x \in [4,5],\\
&= 6-x && \text{for } x \in [5,6],\\
&= 0 && \text{for } x \notin [4,6].
\end{aligned}$$

Then, the membership functions $\mu_{\{i\},\{j,k\}}$ of the sums $w(\{i\})\oplus w(\{j,k\})$, $i,j,k \in I$, $i \neq j \neq k \neq i$, are

$$\begin{aligned}
\mu_{\{i\},\{j,k\}}(x) &= \tfrac{1}{3}x-1 && \text{for } x \in [3,6],\\
&= 4-\tfrac{1}{2}x && \text{for } x \in [6,8],\\
&= 0 && \text{for } x \notin [3,8],
\end{aligned}$$

and the membership function $\mu_{\{1\},\{2\},\{3\}}$ of the sum $w(\{1\})\oplus w(\{2\})\oplus w(\{3\})$ is

$$\begin{aligned}\mu_{\{1\},\{2\},\{3\}}(x) &= \tfrac{1}{3}x-1 \quad \text{for} \quad x\in[3,6],\\ &= 3-\tfrac{1}{3}x \quad \text{for} \quad x\in[6,9],\\ &= 0 \quad \text{for} \quad x\notin[3,9].\end{aligned}$$

Then, it can be easily derived that

$$\begin{aligned}\overline{\nu}(\{i\},\{j,k\}) &= \nu_{\succsim}(w(I),(w(\{i\})\oplus w(\{j,k\}))) = \frac{3}{4} =\\ &= \nu_{\succsim}(w(\{1\})\oplus w(\{2\})\oplus w(\{3\})) = \overline{\nu}(\{1\},\{2\},\{3\}).\end{aligned}$$

since, evidently, $\overline{\nu}(\{i\},\{j\}) = \nu_{\succsim}(w(\{i,j\}),(w(\{i\})\oplus w(\{j\}))) = 1$, for all $i,j\in I$, then

$$\nu^0_{\text{super}}(I,w) = \nu_{\text{super}}(I,w).$$

□

Example 5.4 Let us consider the 3-players game (I,v) described in the previous example but let us modify its fuzzy extension (I,w) in the following way. The values $\mu_{\{i\}}(x)$ remain unchanged, i.e.

$$\begin{aligned}\mu_{\{i\}}(x) &= x-1 \quad \text{for} \quad x\in[1,2],\ i=1,2,3,\\ &= 3-x \quad \text{for} \quad x\in[2,3],\\ &= 0 \quad \text{for} \quad x\notin[1,3],\end{aligned}$$

as well as the values of $\mu_I(x)$, i.e.

$$\begin{aligned}\mu_I(x) &= x-4 \quad \text{for} \quad x\in[4,5],\\ &= 6-x \quad \text{for} \quad x\in[5,6],\\ &= 0 \quad \text{for} \quad x\notin[4,6].\end{aligned}$$

Let us put the membership functions $\mu_{\{i,j\}}$ of $w(\{i,j\})$ as

$$\begin{aligned}\mu_{\{i,j\}}(x) &= x-3 \quad \text{for} \quad x\in[3,4],\ i,j\in I,\ i<j,\\ &= 5-x \quad \text{for} \quad x\in[4,5],\\ &= 0 \quad \text{for} \quad x\notin[3,5].\end{aligned}$$

Then, the membership function $\mu_{\{1\},\{2\},\{3\}}$ remains equal to the one in Example 5.3, and

$$\begin{aligned}\mu_{\{i\},\{j,k\}}(x) &= \tfrac{1}{2}x-2 \quad \text{for} \quad x\in[4,6],\\ &= 4-\tfrac{1}{2}x \quad \text{for} \quad x\in[6,8],\\ &= 0 \quad \text{for} \quad x\notin[4,8].\end{aligned}$$

The possibilities $\overline{\nu}$ are

$$\overline{\nu}(\{i\},\{j,k\}) = \nu_{\succsim}(w(I),(w(\{i\})\oplus w(\{j,k\}))) = \frac{8}{3}$$

and $\overline{\nu}(\{1\},\{2\},\{3\}) = \frac{3}{4}$, also as in Example 5.3.

Since $\overline{\nu}(\{i\},\{j\}) = 1$, for all $i, j \in I$, then

$$\nu^0_{\text{super}}(I, w) = \frac{8}{3}$$

and

$$\nu_{\text{super}}(I, w) = \min\left(\nu^0_{\text{super}}(I, w), \overline{\nu}(\{1\},\{2\},\{3\})\right) = \nu^0_{\text{super}}(I, w).$$

□

Relation (5.5) immediately implies that for any coalition structure $\mathcal{K} = (K_1, \ldots, K_m)$, we have

$$w(I) \succsim w(K_1) \oplus \cdots \oplus w(K_m).$$

Consequently, if an imputation $\boldsymbol{x} = (x_i)_{i \in I}$ is accessible for a coalition structure $\mathcal{K}$ with some possibility, then it is accessible for $w(I)$ with at least the same possibility.

Formally, the possibility that

$$w(K) \succsim \sum\nolimits_K x_i \quad \text{for all } K \in \mathcal{K}$$

is not greater than the possibility of

$$w(I) \succsim \sum\nolimits_I x_i$$

and

$$\nu_{\succsim}\left(w(I), \sum\nolimits_I x_i\right) \geq \min\left[\nu_{\succsim}\left(w(K), \sum\nolimits_K x_i\right) : K \in \mathcal{K}\right].$$

5.2 Fuzzy Subadditivity

The concept of subadditivity in fuzzy extensions of the coalition games with side-payments and its properties are closely analogous to the ones dealt with in the previous section.

Even in this case we can see that the subadditivity based on the relation between pairs of coalition (3.3) does not imply the intuitively more desirable "universal" subadditivity of general m-tuples of coalitions. For example, if (I, v) is a coalition game and (I, w) is its fuzzy extension, $I = \{1, 2, 3\}$, then the inequality $v(K \cup L) \leq v(K) + v(L)$, for any disjoint coalitions $K, L \subset I$, implies

$$v(I) \leq v\left(v(\{1\}) + v(\{2\}) + v(\{3\})\right)$$

but the possibility that for any disjoint $K, L \subset I$

$$w(K) \oplus w(L) \succsim w(K \cup L)$$

need not generally limit from below the possibility of

$$w(\{1\}) \oplus w(\{2\}) + w(\{3\}) \succsim w(I)$$

as illustrated by the following example.

Example 5.5 Let us consider a 3-players game (I, v) with $I = \{1, 2, 3\}$, and

$$v(\{i\}) = 4, \quad v(\{i,j\}) = 6 \text{ for } i,j \in I, \quad v(I) = 14.$$

This game is neither superadditive (as $v(\{i\}) + v(\{j\}) > v(\{i,j\})$, $i, j \in I$, $i \neq j$), nor subadditive (as $v(\{i\}) + v(\{j,k\}) < v(I)$, $i, j, k \in I$, $i \neq j \neq k \neq i$, or $v(\{1\}) + v(\{2\}) + v(\{3\}) < v(I)$).

Let us consider its fuzzzy extension (I, w) with the membership functions $\mu_{\{i\}}$, $\mu_{\{i,j\}}$ and μ_I given by

$$\begin{aligned} \mu_{\{i\}}(x) &= x - 3 && \text{for} \quad x \in [3,4], \ i = 1,2,3, \\ &= 21 - 5x && \text{for} \quad x \in [4, 4.2] \\ &= 0 && \text{for} \quad x \notin [3, 4.2], \end{aligned}$$

$$\begin{aligned} \mu_{\{i,j\}}(x) &= x - 5 && \text{for} \quad x \in [5,6], \ i = 1,2, \ j = 2,3, \ i \neq j, \\ &= \tfrac{8}{5} - \tfrac{1}{10}x && \text{for} \quad x \in [6, 16] \\ &= 0 && \text{for} \quad x \notin [5, 16], \end{aligned}$$

$$\begin{aligned} \mu_I(x) &= x - 13 && \text{for} \quad x \in [13,14], \\ &= 15 - x && \text{for} \quad x \in [14, 15] \\ &= 0 && \text{for} \quad x \notin [13, 15]. \end{aligned}$$

Then, for any triple $i, j, k \in I$, $i \neq j \neq k \neq i$, the possibility that

$$w(\{i\}) \oplus w(\{j,k\}) \succsim w(I)$$

is

$$\nu_{\succsim}((w(\{i\}) \oplus w(\{j,k\})), w(I)) = \frac{9}{14},$$

as the membership function $\mu_{\{i\},\{j,k\}}$ of the sum $w(\{i\}) \oplus w(\{j,k\})$ is

$$\begin{aligned} \mu_{\{i\},\{j,k\}}(x) &= \tfrac{1}{2}x - 4 && \text{for} && x \in [8, 10], \\ &= \tfrac{101}{51} - \tfrac{5}{51}x && \text{for} && x \in [10, 20.2], \\ &= 0 && \text{otherwise.} \end{aligned}$$

The possibility of the relation

$$w(\{1\}) \oplus w(\{2\}) \oplus w(\{3\}) \succsim w(I)$$

is equal to

$$\nu_{\succsim}\left(w(\{1\}) \oplus w(\{2\}) \oplus w(\{3\})\right) w(I) = 0$$

as the membership function $\mu_{\{1\},\{2\},\{3\}}$ of the sum $w(\{1\}) \oplus w(\{2\}) \oplus w(\{3\})$ is

$$\begin{aligned} \mu_{\{1\},\{2\},\{3\}}(x) &= \tfrac{1}{3}x - 3 \quad \text{for} \quad x \in [9, 12], \\ &= 21 - \tfrac{5}{3}x \quad \text{for} \quad x \in [12, 12.6], \\ &= 0 \quad \text{for} \quad x \notin [9, 12.6]. \end{aligned}$$

□

Similarly to the superadditivity concept, we start with a direct two-coalitions analogy to the deterministic case and then show the general concept of fuzzy subadditivity whose general properties are very similar. If (I, w) is a fuzzy extension of (I, v), then we say that is is *simply fuzzy subadditive* iff for any pair of disjoint coalitions K, L there holds

$$w(K) \oplus w(L) \succsim w(K \cup L). \tag{5.7}$$

This fuzzy ordering relation is valid with possibility $\nu_{\succsim}(w(K) \oplus w(L), w(K \cup L))$ given by (2.9) and briefly denoted by $\underline{\nu}(K, L)$, and

$$\nu_{\succsim}(w(K) \oplus w(L), w(K \cup L)) = \underline{\nu}(K, L) = \sup_{\substack{x,y \in R \\ x \geq y}} [\min(\mu_{K+L}(x), \mu_{K \cup L}(y))] \tag{5.8}$$

where the symbol μ_{K+L} denotes the membership function defined by (5.3).

Using (5.8) for all pairs of disjoint coalitions we can derive the possibility that the fuzzy game (I, w) is simply fuzzy subadditive as the number

$$\nu^0_{\text{sub}}(I, w) = \min(\underline{\nu}(K, L) : K, L \subset I, K \cap L = \emptyset). \tag{5.9}$$

It is not difficult to see that, for any pair of disjoint coalitions K, L, $\underline{\nu}(K, L) = \underline{\nu}(L, K)$ and

$$\max(\overline{\nu}(K, L), \underline{\nu}(K, L)) = 1,$$

(see also [28]).

If (I, w) is a fuzzy extension of (I, v) and $K, L \subset I$, $K \cap L = \emptyset$, then evidently $v(K) + v(L) \geq v(K \cup L)$ implies $\underline{\nu}(K, L) = 1$ and, consequently, if (I, v) is subadditive in the deterministic case, then $\nu^0_{\text{sub}}(I, w) = 1$.

Also the possibility of fuzzy subadditivity can be increased if we increase the uncertainty connected with the fuzzy extension. Namely, if (I, w) and (I, w') are fuzzy extensions of (I, v) with membership functions μ_K, μ'_K, for $w(K), w'(K)$, respectively, if $\mu_K(x) \geq \mu'_K(x)$, for all $K \subset I$, $x \in R$, and if we denote for any pair of disjoint coalitions $K, L, \underline{\nu}(K, L)$ and $\underline{\nu}'(K, L)$ the possibility values defined by (5.6) for the respective games, then $\underline{\nu}(K,L) \geq \underline{\nu}'(K,L)$ and, moreover, $\nu^0_{\text{sub}}(I,w) \geq \nu^0_{\text{sub}}(I,w')$.

Analogously to the superadditivity case, this general principle of an increasing possibility of subadditivity with an increasing uncertainty of coalition payments can be extended to its extreme form. For any deterministic game (I,v) there exists its fuzzy extension (I,w) such that, for any pair of disjoint coalitions K, L, $\underline{\nu}(K,L) = 1$ and, consequently, the possibility of subadditivity $\nu^0_{\text{sub}}(I,w) = 1$. On the other hand, this extreme result is possible only if we accept for some coalitions $K \subset I$ the values $\mu_K(x) = 1$ even for $x \neq v(K)$. If (I,w) is a fuzzy extension of (I,v) such that for any $K \subset I$, $\mu_K(x) = 1$ iff $x = v(K)$, then for any pair $K, L \subset I$, $K \cap L = \emptyset$, $\underline{\nu}(K,L) = 1$ if and only if $v(K \cup L) \leq v(K) + v(L)$ and, consequently, $\nu^0_{\text{sub}}(I,w) = 1$ iff (I,v) is subadditive in the deterministic sense.

As (5.7) does not imply the validity of a more general relation

$$w(K_1) \oplus \cdots \oplus w(K_m) \succsim w(K_1 \cup \cdots \cup K_m) \tag{5.10}$$

for a disjoint class of coalitions $\{K_1, \dots, K_m\}$, $K_j \cap K_k = \emptyset$ for $j \neq k$, $j,k = 1, \dots, m$, we can proceed analogously to the superadditivity and denote

$$\underline{\nu}(K_1, \dots, K_m) = \nu_{\succsim}(w(K_1) \oplus \cdots \oplus w(K_m), w(K_1 \cup \cdots \cup K_m)) \tag{5.11}$$

for any class of disjoint coalitions $\{K_1, \dots, K_m\}$.

Then, we denote the possibility that (I,w) is *fuzzy subadditive* by

$$\begin{aligned}\nu_{\text{sub}}(I,w) \quad = \quad & \min\left(\underline{\nu}(K_1, \dots, K_m) : (K_1, \dots, K_m),\ m \geq 2,\ K_j \subset I,\right. \\ & \left. K_j \cap K_k = \emptyset, j \neq k, k, k = 1, \dots, m\right). \end{aligned} \tag{5.12}$$

Relations between the simple fuzzy subadditivity and the more general fuzzy subadditivity (5.10) with possibility (5.12) are closely analogous to those shown for the superadditivity case.

Namely, it is evident that for any fuzzy extension (I,w) of a game (I,v), there is

$$\nu_{\text{sub}}(I,w) \leq \nu^0_{\text{sub}}(I,w).$$

Moreover, other properties of the simple fuzzy subadditivity mentioned above are valid for the fuzzy subadditivity (5.10), too. If (I,w) is a fuzzy extension of (I,v) and if (I,v) is subadditive, then $\nu_{\text{sub}}(I,w) = 1$. For any deterministic game (I,v) there exists its fuzzy extension (I,w) such that $\nu_{\text{sub}}(I,w) = 1$ but if the fuzzy extension does not extend the modal values, i.e. $\mu_K(x) = 1$ iff $x = v(K)$ for all $K \subset I$, then $\nu_{\text{sub}}(I,w) = 1$ if and only if (I,v) is subadditive.

It can also be seen that for, any class of disjoint coalitions $\{K_1, \dots, K_m\}$, there is

$$\max\left(\overline{\nu}(K_1, \dots, K_m), \underline{\nu}(K_1, \dots, K_m)\right) = 1.$$

Example 5.6 Let us consider the coalition game (I, v) and its fuzzy extension (I, w) analyzed in the previous Example 5.5. We have already derived the possibilities of domination

$$\begin{aligned} \underline{\nu}(\{i\}, \{j, k\}) &= \frac{9}{14}, \quad i, j, k \in I, \ i \neq j \neq k \neq i, \\ \underline{\nu}(\{1\}, \{2\}, \{3\}) &= 0. \end{aligned}$$

If we wish to calculate the possibilities $\nu^0_{\text{sub}}(I, w)$ and $\nu_{\text{sub}}(I, w)$, we also need to know the possibilities of the relations $w(\{i\}) \oplus w(\{j\}) \succsim w(\{i, j\})$ for $i, j \in I$, $i \neq j$, i.e. the values $\underline{\nu}(\{i\}, \{j\})$.

It can be easily seen that for any considered pair $i, j \in I$, there holds

$$\underline{\nu}(\{i\}, \{j\}) = 1.$$

Consequently

$$\begin{aligned} \nu^0_{\text{sub}}(I, w) &= \min \Big(\underline{\nu}(\{1\}, \{2\}), \underline{\nu}(\{1\}, \{3\}), \underline{\nu}(\{2\}, \{3\}), \\ &\quad \underline{\nu}(\{1\}, \{2, 3\}), \underline{\nu}(\{2\}, \{1, 3\}), \underline{\nu}(\{3\}, \{1, 2\}) \Big) = \frac{9}{14}. \end{aligned}$$

□

5.3 Fuzzy Additivity

If (I, v) is a deterministic coalition game with side-payments, then its additivity can be defined in two equivalent ways.

First, it is possible to say that (I, v) is both superadditive and subadditive.

Second, we can demand the validity of the following equality

$$v(K \cup L) = v(K) + v(L) \tag{5.13}$$

for any pair of disjoint coalitions $K, L \subset I$.

If the fuzzy extension of (I, v) is considered, then these two approaches are to be treated separately.

Analogously to the deterministic models we will consider the first approach as the primary one.

If (I, w) is a fuzzy extension of a coalition game (I, v), then we say that it is *simply fuzzy additive* iff it is simply fuzzy superadditive and weakly fuzzy subadditive. If $K, L \subset I$, $K \cap L = \emptyset$, then we denote by $\nu(K, L)$ the number

$$\nu(K, L) = \min(\overline{\nu}(K, L), \underline{\nu}(K, L)). \tag{5.14}$$

The possibility that (I,w) is fuzzy additive, denoted by $\nu^0_{\mathrm{addit}}(I,w)$, is the fuzzy logical conjunction of its superadditivity and subadditivity, i.e.

$$\nu^0_{\mathrm{addit}}(I,w) = \min\left(\nu^0_{\mathrm{super}}(I,w), \nu^0_{\mathrm{sub}}(I,W)\right), \qquad (5.15)$$

and it is easy to see that

$$\nu^0_{\mathrm{addit}}(I,w) = \min(\nu(K,L) : K.L \subset I, K \cap L = \emptyset). \qquad (5.16)$$

Even in the case of additivity, $\nu(K,L) = \nu(L,K)$ for any disjoint pair of coalitions.

If (I,w) is a fuzzy extension of a game (I,v) and if (I,v) is additive, then $\nu^0_{\mathrm{addit}}(I,w) = 1$ which follows from analogical properties of superadditivity and subadditivity.

Moreover, the following results follow from their preceeding analogues. If (I,w) and (I,w') are fuzzy extensions of (I,v), if we denote by μ_K and μ'_k the membership functions of w and w', respectively, if $\mu_K(x) \geq \mu'_K(x)$ for any $x \in R$, and if we denote for $K, L \subset I$, $K \cap L = \emptyset$, the values $\nu(K,L)$ and $\nu'(K,L)$ defined by (5.14) for fuzzy games (I,w) and (I,w'), respectively, then $\nu(K,L) \geq \nu'(K,L)$ and

$$\nu^0_{\mathrm{addit}}(I,w) \geq \nu^0_{\mathrm{addit}}(I,w').$$

If (I,v) is a coalition game, then there exists its fuzzy extension (I,w) such that $\nu^0_{\mathrm{addit}}(I,w) = 1$. On the other hand, if for all $K \subset I$ the membership function μ_K fulfills $\mu_K(x) = 1$ iff $x = v(K)$, then $\nu^0_{\mathrm{addit}}(I,w) = 1$ iff (I,v) is additive in the deterministic sense.

It is evident that the approach based on the conjunction of fuzzy superadditivity and fuzzy subadditivity can be quite mechanically used for the general fuzzy additivity, too.

We say that (I,w) is *fuzzy additive* iff it is fuzzy superadditive and fuzzy subadditive, and the possibility that (I,w) is fuzzy additive is

$$\nu_{\mathrm{addit}}(I,w) = \min\left(\nu_{\mathrm{super}}(I,w), \nu_{\mathrm{sub}}(I,w)\right).$$

A close relation to the fuzzy superadditivity and fuzzy subadditivity obviously implies that the general properties of $\nu_{\mathrm{addit}}(I,w)$ be completely analogous to those of $\nu^0_{\mathrm{addit}}(I,w)$.

The illustrative examples presented in the previous two subsections on the fuzzy superadditivity and fuzzy subadditivity can also be used to show that the behaviour of the weak fuzzy additivity and fuzzy additivity, (or, better to say, the behaviour of their particular components). Here, we will present two examples to illustrate this.

Example 5.7 Let us consider the coalition game (I,v) and its fuzzy extension (I,w) already discussed in Examples 5.1 and 5.2.

We already know that

$$\nu^0_{\text{super}}(I,w) = \frac{11}{21}, \quad \nu_{\text{super}}(I,w) = \frac{3}{13}.$$

Let us calculate $\nu^0_{\text{sub}}(I,w)$ and $\nu_{\text{sub}}(I,w)$. It is not difficult to verify that

$$\begin{aligned} &\underline{\nu}(\{i\},\{j\}) = 1, \quad \text{for } i,j \in I,\ i \neq j, \\ &\underline{\nu}(\{1\},\{2,3\}) = 1, \quad \text{for } \underline{\nu}(\{1\},\{2\},\{3\}) = 1. \end{aligned}$$

Moreover, the membership functions $\mu_{\{2\},\{1,3\}}$ and $\mu_{\{3\},\{1,2\}}$ of the sums $w_{\{2\}} \oplus w_{\{1,3\}}$ and $w_{\{3\}} \oplus w_{\{1,2\}}$, respectively, are equal, and

$$\begin{aligned} \mu_{\{2\},\{1,3\}}(x) = \mu_{\{3\},\{1,2\}}(x) &= \tfrac{10}{11}x - \tfrac{89}{11} && \text{for} \quad x \in [8.9, 10], \\ &= 3 - \tfrac{1}{5}x && \text{for} \quad x \in [10, 15], \\ &= 0 && \text{for} \quad x \notin [8.9, 15]. \end{aligned}$$

Hence, the equality also holds between $\underline{\nu}(\{2\},\{1,3\})$ and $\underline{\nu}(\{3\},\{1,2\})$ which are equal to $\frac{5}{6}$. This means that $\underline{\nu}(\{1\},\{2\},\{3\}) = 1$, and

$$\nu^0_{\text{sub}}(I,w) = \nu_{\text{sub}}(I,w) = \frac{5}{6}.$$

Then, the possibilities of additivity are

$$\begin{aligned} \nu^0_{\text{addit}}(I,w) &= \min\left(\nu^0_{\text{super}}(I,w), \nu^0_{\text{sub}}(I,w)\right) = \frac{11}{21}, \\ \nu_{\text{addit}}(I,w) &= \min\left(\nu_{\text{super}}(I,w), \nu_{\text{sub}}(I,w)\right) = \frac{3}{13}. \end{aligned}$$

We can see that the possibility that the fuzzy game (I,w) is subadditive is greater than the possibility that it is superadditive and, consequently, the possibility of superadditivity implies also the possibility of additivity.

□

Example 5.8 Let us return to the coalition game (I,v) and its fuzzy extension (I,w) discussed in Examples 5.5 and 5.6.

We have already investigated its subadditivity and know that

$$\nu^0_{\text{sub}}(I,w) = \frac{9}{14}, \quad \nu_{\text{sub}}(I,w) = 0.$$

Now, we check its supeardditivity. It is easy to verify, using the membership functions of $w(K)$, $K \subset I = \emptyset$, and comparing their modal values, that

$$\begin{aligned}\overline{\nu}(\{i\},\{j,k\}) &= 1 \quad \text{for } i,j,k \in I,\ i \neq j \neq k \neq i,\\ \overline{\nu}(\{1\},\{2\},\{3\}) &= 1.\end{aligned}$$

It is not difficult to compute the possibilities $\overline{\nu}(\{i\},\{j\})$, $i,j \in I$, $i \neq j$, of the relation

$$w(\{i,j\}) \succsim w(\{i\}) \oplus w(\{j\})$$

using the membership function $\mu_{\{i\},\{j\}}$ of the sum $w(\{i\}) \oplus w(\{j\})$, and

$$\begin{aligned}\mu_{\{i\},\{j\}}(x) &= \tfrac{1}{2}x - 3 \quad \text{for} \quad x \in [6,8],\\ &= 21 - \tfrac{5}{2}k \quad \text{for} \quad x \in [8,8.4],\\ &= 0 \quad \text{for} \quad x \notin [6,8.4].\end{aligned}$$

We can easily see that $\overline{\nu}(\{i\},\{j\}) = \frac{5}{6}$.

Then

$$\begin{aligned}\nu^0_{\text{super}}(I,w) &= \min\Big(\overline{\nu}(\{1\},\{2\}),\overline{\nu}(\{1\},\{3\}),\overline{\nu}(\{2\},\{3\}),\\ &\qquad \overline{\nu}(\{1\},\{2,3\}),\overline{\nu}(\{2\},\{1,3\}),\overline{\nu}(\{3\},\{1,2\})\Big) = \frac{5}{6},\end{aligned}$$

and also $\nu_{\text{super}}(I,w) = \frac{5}{6}$ as $\overline{\nu}(\{1\},\{2\},\{3\}) = 1$.

It means that the possibilities of additivity are

$$\begin{aligned}\nu^0_{\text{addit}}(I,w) &= \min\left(\nu^0_{\text{super}}(I,w),\nu^0_{\text{sub}}(I,w)\right) = \frac{9}{14},\\ \nu_{\text{addit}}(I,w) &= \min\left(\nu_{\text{super}}(I,w),\nu_{\text{sub}}(I,w)\right) = 0.\end{aligned}$$

The structure of the fuzzy coalition game (I,w) considered in this example is nearer to the superadditivity than to the subadditivity and the possibilities of the subadditivity, in fact, determine the possibility of the additivity.

□

If we intend to fuzzify the second approach to the additivity based on equality (5.13), then we have to use a generalized fuzzy analogue of "equality" (2.10) and define for any set of coalitions $\{K_1,\ldots,K_m\}$, $K_j \cap K_k = \emptyset$, for $j \neq k$, $j,k = 1,\ldots,m$, the possibility

$$\nu^*(K_1,\ldots,K_m) = \nu_{\sim}\left(w(K_1 \cup \cdots \cup K_m), w(K_1) \oplus \cdots \oplus w(K_m)\right), \tag{5.17}$$

where (2.10) was used.

Using (5.17) we define the value

$$\begin{aligned}\nu^*_{\text{addit}}(I,w) &= \min\left(\nu^*(K_1,\ldots,K_m) : (K_1,\ldots,K_m), m \geq 2, K_j \subset I,\right.\\ &\qquad \left. K_j \cap K_k = \emptyset, j \neq k, j,k = 1,\ldots,m\right).\end{aligned} \tag{5.18}$$

In the introduction of this part we have presented assumptions on the membership functions μ_K, among them also their monotonicity (not necessarily strict); $\mu_K(x)$ is nondecreasing for $x < v(K)$ and nonincreasing for $x > v(K)$. This means that for any m-tuple of disjoint coalitions $\{K_1, \ldots, K_m\}$, also the membership function of the sum $w(K_1) \oplus \cdots \oplus w(K_m)$ has the same monotonicity property. It means that, for any such disjoint m-tuple of coalitions, we have

$$\nu^*(K_1, \ldots, K_m) = \min \left(\overline{\nu}(K_1, \ldots, K_m), \underline{\nu}(K_1, \ldots, K_m)\right)$$

where at least one of the the possibilities $\overline{\nu}(K_1, \ldots, K_m)$ and $\underline{\nu}(K_1, \ldots, K_m)$ is equal to 1.

The previous equality means that

$$\nu^*_{\text{addit}}(I, w) = \nu_{\text{addit}}(I, w)$$

and, consequently, $\nu^*_{\text{addit}}(I, w)$ and $\nu^*(K_1, \ldots, K_m)$ have the properties already described for $\nu_{\text{addit}}(I, w)$ and $\nu(K_1, \ldots, K_m)$.

Moreover, this concept of additivity can be formulated in a "weak" form based on a pair-wise relation between disjoint coalitions. The previous subsection illustrates however an intuitively expectable conclusion that the properties of such a weak fuzzy additivity are completely analogous to the general ones.

6. Fuzzy Core and Effective Coalitions

The concept of a core in a fuzzy extension of a coalition game with side-payments will be naturally based on fuzzy analogues of (3.5) and (3.6). A heuristic analysis of the investigated model leads to the conclusion that a core of a fuzzified game is to be fuzzified as well. The vagueness entering the game by means of uncertain expectations of coalitional profits is also reflected by the vagueness connected with outputs of the game including the concept of a core. In the deterministic game, the core C is a subset of R^I. Hence, in the fuzzy extension it should also be a fuzzy subset of the same set, as shown, e.g., in [33].

Namely, if (I, v) is a coalition game with side-payments and (I, w) is its fuzzy extension, then the *fuzzy core* C_F *of* (I, w) is a fuzzy subset C_F of R^I with the membership function $\gamma_C : R^I \to [0, 1]$.

For any $\boldsymbol{x} \in R^I$ the value $\gamma_C(\boldsymbol{x})$ is the possibility that $\boldsymbol{x}$ is an element of the core, i.e. that it is achievable by some coalition structure and it cannot be protested by any coalition. In a more formal language, if $\mathcal{K}$ is a coalition structure, then the possibility that $\boldsymbol{x}$ is achievable by $\mathcal{K}$ is

$$\begin{aligned} \tau_{\mathcal{K}}(\boldsymbol{x}) &= \min\left(\nu_{\succsim}\left(w(K), \sum\nolimits_{K} x_i\right) : K \in \mathcal{K}\right) = \\ &= \min\left(\sup\left(\mu_K(y) : y \in R, y \geq \sum\nolimits_{K} x_i\right) : K \in \mathcal{K}\right) \end{aligned}$$

which follows from (2.12). This means that the possibility that $\boldsymbol{x}$ is achievable for at least one coalition structure is

$$\tau_C(\boldsymbol{x}) = \max\left(\tau_{\mathcal{K}}(\boldsymbol{x}) : \mathcal{K} \text{ is a coalition structure}\right).$$

On the other hand, the possibility that $\boldsymbol{x}$ cannot be protested by a coalition $L \subset I$ is the possibility that

$$\sum\nolimits_{L} x_i \succsim w(L),$$

which we denote, using (2.11), as

$$\nu_{\succsim}\left(\sum\nolimits_{L} x_i, w(L)\right) = \sup\left(\mu_L(y) : y \in R, y \leq \sum\nolimits_{L} x_i\right),$$

and the possibility that $\boldsymbol{x}$ is not protested by any $L \subset I$ is

$$\lambda_C(\boldsymbol{x}) = \min\left(\nu_{\succsim}\left(\sum\nolimits_L x_i, w(L)\right) : L \subset I\right).$$

Then the possibility $\gamma_C(\boldsymbol{x})$ is given by

$$\gamma_C(\boldsymbol{x}) = \min(\tau_C(\boldsymbol{x}), \lambda_C(\boldsymbol{x})). \tag{6.1}$$

The fuzzy core preserves some properties of the classical deterministic one. For example, if (I, v) is superadditive, then – as we already know – its fuzzy extension (I, w) is fuzzy superadditive and, for any coalition structure $\mathcal{K}$, we have

$$\tau_{\mathcal{K}}(\boldsymbol{x}) \leq \nu_{\succsim}\left(w(I), \sum\nolimits_I x_i\right) = \tau_{\{I\}}(\boldsymbol{x})$$

where $\{I\}$ is the coalition structure containing exactly one coalition I.

Then, consequently,

$$\gamma_C(\boldsymbol{x}) = \min\left(\nu_{\succsim}\left(w(I), \sum_I x_i\right), \lambda_C(\boldsymbol{x})\right). \tag{6.2}$$

There exist relations between the deterministic core C of the game (I, v) and the fuzzy core C_F of its fuzzy extension (I, w), analogous to some results already presented for additivity-like concepts.

If (I, w) is a fuzzy extension of (I, v), if C is the core of (I, v) and if $\boldsymbol{x} \in C$, then the properties of a fuzzy ordering relation $\succsim$ imply that $\gamma_C(\boldsymbol{x}) = 1$. It means that the fuzzy core C_F of (I, w) is a fuzzy extension of its deterministic counterpart C as shown in [33].

On the other hand, for every coalition game (I, v) there exists such an extension (I, w) that $\gamma_C(\boldsymbol{x}) = 1$, for some $\boldsymbol{x} \in R^I$. It is sufficient to define $w(I)$ "wide" enough. But, if for every $L \subset I$, $\mu_L(x) < 1$ for $\boldsymbol{x} \neq v(L)$, then

$$\{\boldsymbol{x} \in R^I : \gamma_C(\boldsymbol{x}) = 1\} = C$$

as also shown in [33].

This result, rather analogous to some properties shown in the case of superadditivity and related topics, confirms the intuitive meaning that if the fuzzy extension of a game does not extend the "certain" values of expected pay-offs, then it cannot influence the "certain" values of rational results of bargaining.

Having introduced the possibility $\gamma_C(\boldsymbol{x})$ that an imputation $\boldsymbol{x} \in R^I$ belongs to the fuzzy core C_F, we can derive the possibility that C_F is non-empty being a number $\gamma(I, w) \in [0, 1]$ defined by

$$\gamma(I, w) = \sup(\gamma_C(\boldsymbol{x}) : \boldsymbol{x} \in R^I). \tag{6.3}$$

The previously shown properties of the membership function γ imply that $\gamma(I, w) = 1$ if the deterministic core C of (I, v) is non-empty and, moreover, if for all coalitions $K \subset I$, $\mu_K(\boldsymbol{x}) = 1$ iff $\boldsymbol{x} = v(K)$, then $\gamma(I,w) = 1$ iff $C \neq \emptyset$.

The conditions for the existence of a non-empty core derived in the deterministic game model, namely their convexity and balancedness, can be used for the investigation of a fuzzy core and its possibility. It is done in the following chapters.

The previous concepts and their properties can be illustrated by the following examples.

Example 6.1 Let us consider a simple 2-players coalition game (I,v) where $I = \{1,2\}$, $v(\{1\}) = 1$, $v(\{2\}) = 2$, $v(I) = 5$. The game is evidently superadditive (and, of course, convex and balanced), and its non-empty core C is the set of imputations representing the segment between $(1,4)$ and $(3,2)$, i.e.

$$C = \left\{\boldsymbol{x} \in R^2 : x_1 + x_2 = 5,\ x_1 \geq 1,\ x_2 \geq 2\right\},$$

which is attainable for the coalition of all players.

Let us consider the fuzzy extension (I,w) of (I,v), where

$$\begin{aligned}
\mu_{\{1\}}(x) &= x && \text{for} && x \in [0,1],\\
&= 2-x && \text{for} && x \in [1,2],\\
&= 0 && \text{otherwise,}
\end{aligned}$$

$$\begin{aligned}
\mu_{\{2\}}(x) &= x-1 && \text{for} && x \in [1,2],\\
&= 3-x && \text{for} && x \in [2,3],\\
&= 0 && \text{otherwise,}
\end{aligned}$$

$$\begin{aligned}
\mu_I(x) &= x-4 && \text{for} && x \in [4,5],\\
&= 6-x && \text{for} && x \in [5,6],\\
&= 0 && \text{otherwise.}
\end{aligned}$$

There exist exactly two coalition structures, namely (I) and $(\{1\},\{2\})$, in this game. It means that the weak superadditivity, subadditivity and additivity have to be identical with their generalized modifications, i.e.

$$\begin{aligned}
\nu^0_{\text{super}}(I,w) &= \nu_{\text{super}}(I,w) = 1,\\
\nu^0_{\text{sub}}(I,w) &= \nu_{\text{sub}}(I,w) = \frac{1}{3},\\
\nu^0_{\text{addit}}(I,w) &= \nu_{\text{addit}}(I,w) = \frac{1}{3}.
\end{aligned}$$

By computing the fuzzy core C_F we can easily see that for the coalition structure $\mathcal{K} = (I)$, we have

$$\tau_{\mathcal{K}}(\boldsymbol{x}) = \tau_{(I)}(\boldsymbol{x}) = \max\left(0, \min(1, 6-(x_1+x_2))\right),$$

which means that for any $\boldsymbol{x} = (x_1,x_2) \in R^2$

$$\begin{aligned}\tau_{(I)}(\boldsymbol{x}) &= 1 && \text{for} \quad x_1 + x_2 \le 5,\\ &= 6 - (x_1 + x_2), && \text{for} \quad 5 < x_1 + x_2 \le 6,\\ &= 0, && \text{for} \quad 6 < x_1 + x_2.\end{aligned}$$

Analogously, for the coalition structure $\mathcal{L} = (\{1\}, \{2\})$, there holds

$$\begin{aligned}\tau_{\mathcal{L}}(\boldsymbol{x}) &= \tau_{(\{1\},\{2\})}(\boldsymbol{x}) = \max\left(0, \min(1, 2 - x_1, 3 - x_2)\right) = \\ &= \min\left(\nu_{\succsim}(w(\{1\}), x_1), \nu_{\succsim}(w(\{2\}), x_2)\right).\end{aligned}$$

This means, in our case, that for $\boldsymbol{x} = (x_1, x_2) \in R^2$

$$\begin{aligned}\tau_{\mathcal{L}}(\boldsymbol{x}) &= 1 && \text{for} \quad x_1 \le 1 \text{ and } x_2 \le 2,\\ &= 2 - x_1 && \text{for} \quad 1 < x_1 \le 2 \text{ and } x_2 \le x_1 + 1,\\ &= 3 - x_2 && \text{for} \quad 2 < x_2 \le 3 \text{ and } x_1 \le x_2 - 1,\\ &= 0 && \text{for} \quad x_1 > 2 \text{ or } x_2 > 3.\end{aligned}$$

It is evident that the function τ_C is determined by $\tau_{\mathcal{K}}$ as $\tau_C(\boldsymbol{x}) = \min(\tau_{\mathcal{K}}(\boldsymbol{x}), \tau_{\mathcal{L}}(\boldsymbol{x})) = \tau_{\mathcal{K}}(\boldsymbol{x})$, for all $\boldsymbol{x} \in R^2$.

Function λ_C can be computed similarly as in the previous procedure. Then, we derive for any $\boldsymbol{x} = (x_1, x_2) \in R^2$

$$\lambda_C(\boldsymbol{x}) = \max\left(0, \min(1, x_1, x_2 - 1, x_1 + x_2 - 4)\right),$$

which means that

$$\begin{aligned}\lambda_C(\boldsymbol{x}) &= 1 && \text{for} \quad x_1 + x_2 \ge 5 \text{ and } x_1 \ge 1 \text{ and } x_2 \ge 2,\\ &= x_1 && \text{for} \quad 0 \le x_1 < 1 \text{ and } x_2 \ge 4,\\ &= x_2 - 1 && \text{for} \quad 1 \le x_2 < 2 \text{ and } x_1 \ge 3,\\ &= x_1 + x_2 - 4 && \text{for} \quad 4 \le x_1 + x_2 < 5 \text{ and } x_1 \ge 0 \text{ and } x_2 \ge 1,\\ &= 0 && \text{for} \quad x_1 + x_2 < 4 \text{ or } x_1 < 0 \text{ or } x_2 < 1.\end{aligned}$$

Having computed the functions τ_C and λ_C, we can easily compute the membership function γ_C of the fuzzy core C_F as their minimum

$$\begin{aligned}\gamma_C(\boldsymbol{x}) &= \min\left(\tau_C(\boldsymbol{x}), \lambda_C(\boldsymbol{x})\right) = \\ &= \max\left(0, \min(6 - (x_1 + x_2), x_1, x_2 - 1, x_1 + x_2 - 4)\right).\end{aligned}$$

In another analytic representation, for all $\boldsymbol{x} = (x_1, x_2) \in R^2$,

$$
\begin{aligned}
\gamma_C(\boldsymbol{x}) &= 6-(x_1+x_2) && \text{for } 5 \le x_1+x_2 \le 6 \\
&&& \text{and } x_1 \ge 0 \text{ and } x_2 \ge 1 \\
&&& \text{and } x_1+2x_1 \ge 7 \\
&&& \text{and } 2x_1+x_2 \ge 6, \\
&= x_1+x_2-4 && \text{for } 4 \le x_1+x_2 \le 5 \text{ and} \\
&&& x_1 \ge 0 \text{ and } x_2 \ge 1, \\
&= x_1 && \text{for } 0 \le x_1 \le 1, \text{ and } 4 \le x_2 \\
&&& \text{and and } x_2 \le 6-2x_1, \\
&= x_2-1 && \text{for } 1 \le x_2 \le 2 \text{ and } 3 \le x_1 \\
&&& \text{and and } x_1 \le 7-2x_2, \\
&= 0 && \text{for } x_1+x_2>6 \\
&&& \text{or } x_1+x_2<5, \text{ or} \\
&&& \text{or } x_1<0 \text{ or } x_2<1.
\end{aligned}
$$

It can be verified that, really, $\gamma_C(\boldsymbol{x}) = 1$ if and only if $\boldsymbol{x} = (x_1, x_2)$ belongs to the segment between $(1,4)$ and $(3,2)$, i.e. iff $\boldsymbol{x} \in C$, and that there exists a positive possibility to be an element of the core for imputations $\boldsymbol{x}$ from the trapezoid $(3,1), (5,1), (0,4), (0.6)$ (see Figure 5.1).

□

Example 6.2 Let us consider a 2-players coalition game (I, v) with $I = \{1,2\}, v(\{1\}) = v(\{2\}) = 6,\ v(I) = 9$. The game is subadditive with non-empty one-element core

$$C = \{(6,6)\}$$

which is realized by the coalition structure $\mathcal{L} = (\{1\},\{2\})$. It is neither superadditive nor convex. It is also not ballanced which implies the fact that the core is not achievable for the all-players coalition I.

Let us consider the fuzzy extension (I, w) of (I, v) with

$$
\begin{aligned}
\mu_{\{1\}}(x) = \mu_{\{2\}}(x) &= x-5 && \text{for} && x \in [5,6], \\
&= 7-x && \text{for} && x \in [6,7], \\
&= 0 && \text{otherwise,}
\end{aligned}
$$

$$
\begin{aligned}
\mu_I(x) &= x-8 && \text{for} && x \in [8,9], \\
&= 10-x && \text{for} && x \in [9,10], \\
&= 0 && \text{otherwise.}
\end{aligned}
$$

The fuzzy coalition game (I, w) is fuzzy subadditive and weakly fuzzy subadditive with possibility

$$\nu^0_{\text{sub}}(I,w) = \nu_{\text{sub}}(I,w) = 1$$

and its possibility, being fuzzy superadditive, vanishes, i.e.

$$\nu^0_{\text{super}}(I,w) = \nu_{\text{super}}(I,w) = 0.$$

Consequently, $\nu^0_{\text{addit}}(I,w) = \nu_{\text{addit}}(I,w) = 0$.

Then, for the coalition structure $\mathcal{K} = (I)$ formed by the single coalition of all players, and $\boldsymbol{x} = (x_1, x_2) \in R^2$, we have

$$\begin{aligned}
\tau_{\mathcal{K}}(\boldsymbol{x}) &= \max(0, \min(1, 10 - (x_1 + x_2))) = \\
&= 1 && \text{for } x_1 + x_2 \leq 9, \\
&= 10 - (x_1 + x_2) && \text{for } 9 < x_1 + x_2 \leq 10, \\
&= 0 && \text{for } x_1 + x_2 > 10.
\end{aligned}$$

For the coalition structure $\mathcal{L} = (\{1\}, \{2\})$, $\boldsymbol{x} = (x_1, x_2) \in R^2$, there holds

$$\begin{aligned}
\tau_{\mathcal{L}}(\boldsymbol{x}) &= \max(0, \min(1, 7 - x_1, 7 - x_2)) = \\
&= 1 && \text{for } x_1 \leq 6 \text{ and } x_2 \leq 6, \\
&= 7 - x_1 && \text{for } 6 < x_1 \leq 7,\ x_2 \leq x_1, \\
&= 7 - x_2 && \text{for } 6 < x_2 \leq 7,\ x_1 \leq x_2, \\
&= 0 && \text{for } x_1 > 7 \text{ or } x_2 > 7.
\end{aligned}$$

The membership function $\tau_C(\boldsymbol{x})$ fulfills

$$\begin{aligned}
\tau_C(\boldsymbol{x}) &= \max(\tau_{\mathcal{K}}(\boldsymbol{x}), \tau_{\mathcal{L}}(\boldsymbol{x})) = \\
&= \max(0, \min(1, 7 - x_1, 7 - x_2)), \min(1, 10 - (x_1 + x_2)) = \\
&= 1 && \text{for } x_1 + x_2 \leq 9 \text{ or } (x_1 \leq 6 \text{ and } x_2 \leq 6), \\
&= 10 - (x_1 + x_2) && \text{for } 9 < x_1 + x_2 \leq 10 \text{ and } (x_1 \leq 3 \text{ or } x_2 \leq 3), \\
&= 7 - x_1 && \text{for } 6 < x_1 \leq 7 \text{ and } 3 \leq x_2 \leq x_1, \\
&= 7 - x_2 && \text{for } 6 < x_2 \leq 7 \text{ and } 3 \leq x_1 \leq x_2, \\
&= 0 && \text{for } x_1 + x_2 > 7 \text{ and } (x_1 > 7 \text{ or } x_2 > 7).
\end{aligned}$$

The computation of the membership function λ_C is similar. For $\boldsymbol{x} = (x_1, x_2) \in R^2$

$$\begin{aligned}
\lambda_C(\boldsymbol{x}) &= \max(0, \min(1, x_1 - 5, x_2 - 5)) = \\
&= 1 && \text{for } x_1 \geq 6 \text{ and } x_2 \geq 6, \\
&= x_1 - 5 && \text{for } 5 \leq x_1 < 6 \text{ and } x_2 \geq x_1, \\
&= x_2 - 5 && \text{for } 5 \leq x_2 < 6 \text{ and } x_1 \geq x_2, \\
&= 0 && \text{for } x_1 < 5 \text{ or } x_2 < 5.
\end{aligned}$$

Finally, the membership function γ_C of the core C_F is for every $\boldsymbol{x} = (x_1, x_2)$ given by

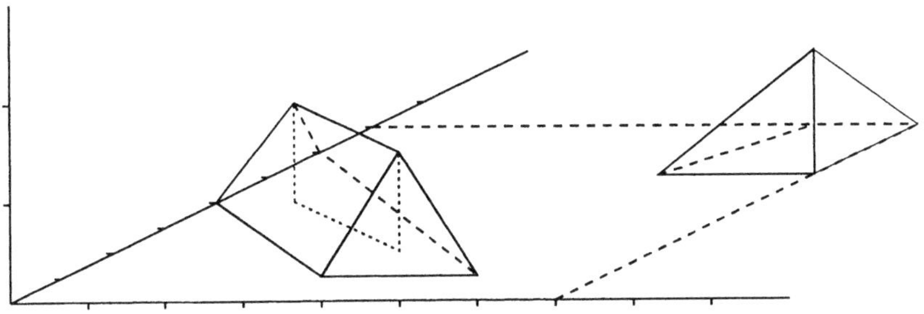

Fig. 6.1. Fuzzy cores (membership functions γ_C) of games in Example 6.1 (C1) and Example 6.2 (C2)

$$\begin{aligned}
\gamma_C(\boldsymbol{x}) &= \min(\tau_C(\boldsymbol{x}), \lambda_C(\boldsymbol{x})) = \\
&= \max(0, \min(1, 7 - x_1, 7 - x_2, x_1 - 5, x_2 - 5)) = \\
&= x_1 - 5 \quad \text{for} \quad 5 \le x_1 \le 6 \text{ and } x_1 \le x_2 \le 12 - x_1, \\
&= 7 - x_1 \quad \text{for} \quad 6 \le x_1 \le 7 \text{ and } 12 - x_1 \le x_2 \le x_1, \\
&= 7 - x_2 \quad \text{for} \quad 6 \le x_2 \le 7 \text{ and } 12 - x_2 \le x_1 \le x_2, \\
&= 0 \quad \text{for} \quad x_1 < 5 \text{ or } x_1 > 7 \text{ or } x_2 < 5 \text{ or } x_2 < 7.
\end{aligned}$$

It means that only the point $(6,6)$ belongs to the fuzzy core C_F with possibility $\gamma_C(\boldsymbol{x}) = 1$ and any imputation outside the square $(5,5)$, $(5,7)$, $(7,7)$, $(7,5)$ certainly does not belong to C_F. The possibility of the existence of a non-empty core is $\gamma(I, w) = 1$ (see Figure 6).

□

The coalition games of 3 and more players are not as clear as the previous ones. In the following example we present one of them.

Example 6.3 Let us fuzzify the 3-players coalition game (I, v) presented in Example 3.2 such that $I = \{1,2,3\}$, $v(\{i\}) = 0$, $v(\{i,j\}) = 2$, for $i,j \in I$, $i \neq j$, and $v(I) = 2$. This game is superadditive but not convex as

$$v(\{1,2\}) + v(\{1,3\}) > v(I) + v(\{1\}),$$

and it is not subadditive. Its deterministic core C is empty.

Let us consider its fuzzy extension (I, w) with

$$\begin{array}{llll} \mu_{\{i\}}(x) & = & x+1 & \text{for} \quad x \in [-1,0],\ i = 1,2,3, \\ & = & 1-x & \text{for} \quad x \in [0,1], \\ & = & 0 & \text{for} \quad x \notin [-1,1], \end{array}$$

$$\begin{array}{llll} \mu_{\{i,j\}}(x) & = & x-1 & \text{for} \quad x \in [1,2],\ i,j \in I,\ i \neq j, \\ & = & 3-x & \text{for} \quad x \in [2,3], \\ & = & 0 & \text{for} \quad x \notin [1,3], \end{array}$$

$$\mu_I(x) \quad = \quad \mu_{\{i,j\}}(x), \quad \text{for} \quad x \in R.$$

We can briefly mention that the fuzzy coalition game (I, w) is certainly superadditive in both of the above senses, i.e.

$$\nu^0_{\text{super}}(I, w) = \nu_{\text{super}}(I, w) = 1,$$

and that it is to a certain small degree also subadditive, i.e.

$$\nu^0_{\text{sub}}(I, w) = \nu_{\text{sub}}(I, w) = \frac{1}{3}.$$

Investigating the properties of its core, we can, first, notice that there exist the following coalition structures

$$(I),\ (\{1\},\{2,3\}),\ (\{2\},\{1,3\}),\ (\{3\},\{1,2\}),\ (\{1\},\{2\},\{3\}),$$

and that the symmetry of the fuzzzy characteristic function values $w(K)$ can essentially simplify the determination of the fuzzy core.

For every imputation $\boldsymbol{x} = (x_1, x_2, x_3) \in R^3$ we have

$$\begin{array}{lll} \tau_{(I)}(\boldsymbol{x}) & = & \max\left(0, \min(1, 3 - (x_1 + x_2 + x_3))\right) = \\ & = & 1 \qquad\qquad\qquad\quad \text{for} \quad x_1 + x_2 + x_3 \leq 2, \\ & = & 3 - (x_1 + x_2 + x_3) \quad \text{for} \quad 2 < x_1 + x_2 + x_3 \leq 3, \\ & = & 0 \qquad\qquad\qquad\quad \text{for} \quad x_1 + x_2 + x_3 > 3, \end{array}$$

$$\begin{array}{lll} \tau_{(\{i\},\{j,k\})} & = & \max\left(0, \min(1, 1 - x_i, 3 - (x_j + x_k))\right), \\ & & \qquad i, j, k \in I,\ i \neq j \neq k \neq i, \end{array}$$

$$\tau_{(\{1\},\{2\},\{3\})} \quad = \quad \max\left(0, \min(1, 1 - x_1, 1 - x_2, 1 - x_3)\right).$$

It means that

$$\begin{aligned}
&\tau_C(\boldsymbol{x}) = \\
&= \max(\tau_{(I)}(\boldsymbol{x}), \tau_{(\{1\},\{2,3\})}(\boldsymbol{x}), \tau_{(\{2\},\{1,3\})}(\boldsymbol{x}), \\
&\qquad \tau_{(\{1\},\{2,3\})}(\boldsymbol{x}), \tau_{(\{3\},\{1,2\})}(\boldsymbol{x})) = \\
&= \max\Big[0, \min(1, 3-(x_1+x_2+x_3)), \min(1, 1-x_1, 3-(x_2+x_3)), \\
&\qquad \min(1, 1-x_2, 3-(x_1+x_3)), \min(1, 1-x_3, 3-(x_1+x_2)), \\
&\qquad \min(1, 1-x_1, 1-x_2, 1-x_3)\Big] = \\
&= \max\Big[0, \min(1, 3-(x_1+x_2)), \min(1, 3-(x_2+x_3)), \\
&\qquad \min(1, 3-(x_1+x_3)), \min(1, 3-(x_1+x_2+x_3))\Big].
\end{aligned}$$

Analogously, we have

$$\begin{aligned}
\lambda_C(\boldsymbol{x}) &= \max\Big[0, \min(1, x_1+x_2-1, x_1+x_3-1, x_2+x_3-1, \\
&\qquad x_1+x_2+x_3-1, x_1+1, x_2+1, x_3+1)\Big]
\end{aligned}$$

and

$$\gamma_C(\boldsymbol{x}) = \min(\tau_C(\boldsymbol{x}), \lambda_C(\boldsymbol{x})).$$

It is possible to verify that $\gamma_C(\boldsymbol{x}) < 1$ for all $\boldsymbol{x} \in R^3$ and $\gamma_C(\boldsymbol{x}) = 0$ for $\boldsymbol{x}$ not belonging to the cube between the vertices:

$$\begin{gathered}
\left(\tfrac{1}{2}, \tfrac{1}{2}, \tfrac{1}{2}\right), \\
\left(1, \tfrac{1}{2}, \tfrac{1}{2}\right), \\
\left(\tfrac{1}{2}, 1, \tfrac{1}{2}\right), \\
\left(\tfrac{1}{2}, \tfrac{1}{2}, 1\right), \\
\left(1, 1, \tfrac{1}{2}\right), \\
\left(1, \tfrac{1}{2}, 1\right), \\
\left(\tfrac{1}{2}, 1, 1\right), \\
(1, 1, 1),
\end{gathered}$$

and its maximal value is attained for the imputation $\left(\frac{4}{5}, \frac{4}{5}, \frac{4}{5}\right)$ for which

$$\gamma_C\left(\frac{4}{5}, \frac{4}{5}, \frac{4}{5}\right) = \frac{3}{5}.$$

This means that also the possibility of the existence of non-empty core is equal to

$$\gamma(I, w) = \frac{3}{5}.$$

The analytic form of the membership function $\gamma_C(\boldsymbol{x})$ is rather complicated but some of its values can be mentioned for illustration. The function $\gamma_C(\boldsymbol{x})$ is continuous which means that $\gamma_C(\boldsymbol{x}) = 0$ also on the vertices, edges and the surface of the cube mentioned above.

Some other values are

$$\gamma_C\left(\frac{3}{4},\frac{3}{4},\frac{3}{4}\right)=\frac{1}{2},\quad \gamma_C\left(\frac{2}{3},\frac{2}{3},\frac{2}{3}\right)=\frac{1}{3},$$

$$\gamma_C\left(\frac{5}{6},\frac{5}{6},\frac{5}{6}\right)=\frac{1}{2},\quad \gamma_C\left(\frac{3}{5},\frac{3}{5},\frac{3}{5}\right)=\frac{1}{5},$$

$$\gamma_C\left(\frac{3}{5},\frac{9}{10},\frac{9}{10}\right)=\frac{1}{10},\quad \gamma_C\left(\frac{3}{5},\frac{3}{5},\frac{3}{4}\right)=\frac{1}{5},\quad \gamma_C\left(\frac{3}{5},\frac{3}{5},\frac{9}{10}\right)=\frac{1}{5},$$

etc.

The previous properties of (I, w) show that even if the deterministic core C is empty, its fuzzy counterpart C_F can contain some possible but not certain values. A further analysis of this game can be found even in Examples 6.6, 7.1, 7.2, 7.4.

□

The core of a coalition game characterizes the distributions of utility which can appear as a result of bargaining. In the deterministic case it only determines the achievable and safe imputations, in the fuzzy case it also structures the degree of their possibility. In fact, the imputations are not a single result of barganing.

Another, also important, outcome of the process of negotiation is the realization of some coalitions and coalition structures. Evidently, some coalitions can be formed while others are excluded. In the case of fuzzy coalition games, moreover, the possible coalitions can be structured with respect to the possibility of their forming. The ability of a coalition to appear as a result or a semi-result of negotiation is also called its effectivity (see, e.g., [29]).

A heuristic characterization of an effective coalition rather remembers the specification of the core – more exactly, its conceptual complement. It emphasises some phenomena which are also respected in the specification of a core but which are, finally, rather suppressed when a core is formally defined.

A coalition is effective if it can offer such a distribution of its coalitional income which cannot be improved by any of its subcoalitions. In the case of a fuzzy coalition game we can also consider the possibility that a coalition is effective as a possibility that it can form such an imputation.

In formal expressions, we say that a coalition $K \subset I$ is effective if there exists an imputation $\boldsymbol{x}_K = (x_i)_{i \in K}$ which is accessible, i.e.

$$w(K) \succsim \sum\nolimits_K x_i \tag{6.4}$$

and if no subcoalition $J \subset K$ can effectively object against $\boldsymbol{x}_K$, i.e.

$$\sum_J x_i \succsim w(J) \quad \text{for all } J \subset K. \tag{6.5}$$

In a fuzzy coalition game both the required properties (6.4) and (6.5) are fuzzy properties valid with some possibility and, consequently, also the property of a coalition K to be effective is a fuzzy property. It is valid with a possibility described by the membership function $\varepsilon : 2^I \to [0,1]$. Its values are derived from the possibilities of (6.4) and (6.5) by

$$\varepsilon(K) =$$
$$= \sup_{\boldsymbol{x}_K \in R^K} \left[\min \left(\nu_{\succsim} \left(w(K), \sum_K x_i \right), \min_{J \subset K} \left(\nu_{\succsim} \left(\sum_J x_i, w(J) \right) \right) \right) \right]. \tag{6.6}$$

It can easily be shown that for any coalition $K \subset I$ which is effective in the deterministic game (I, v), the possibility $\varepsilon(K) = 1$. For every game (I, v) there exists its fuzzy extension (I, w) in which all coalitions are certainly effective (i.e. $\varepsilon(K) = 1$ for all $K \subset I$) but if an extension (I, w) fulfills $\mu_K(x) = 1$ iff $x = v(K)$ for all coalitions K, then $\varepsilon(K) = 1$ iff K is effective in the deterministic game (I, v).

A close relation between the concepts of a core and effectivity for superadditive games is evident. The coalition of all players I is effective with the possibility

$$\varepsilon(I) = \sup(\gamma_C(\boldsymbol{x}) : \boldsymbol{x} \in R^I) = \gamma(I, w).$$

Consequently, in the superadditive games, $\varepsilon(I) = 1$ if the deterministic core C of (I, v) is non-empty.

The concepts of effectivity and core describe possible results of bargaining – the possibility of stable imputations and the possibility of forming stable coalitions.

The effectivity of coalitions can be illustrated by the games presented already in the previous examples. We can see that, of course, all one-element coalitions $\{i\}$, $i \in I$, are certainly effective, i.e. $\varepsilon(\{i\}) = 1$.

Example 6.4 The effectivity in 2-players games is very simple. Let us consider the fuzzy coalition game investigated in Example 6.1. Then, evidently, $\varepsilon(\{i\}) = 1$ for $i = 1, 2$, and $\varepsilon(I) = 1$, as well, as, e.g., for $\boldsymbol{x} = (1, 4)$, there is

$$\nu_{\succsim}(w(I), 5) = 1, \quad \nu_{\succsim}(1, w(\{1\})) = 1,, \quad \nu_{\succsim}(4, w(\{2\})) = 1.$$

□

Example 6.5 The effectivity in the fuzzy coalition game investigated in Example 6.2 is not as self-evident as the same property in the previous example. Of course, $\varepsilon(\{1\}) = 1$ as for $x_1 = 6$ we have

$$\nu_{\succsim}(w(\{1\}), 6) = \nu_{\succsim}(6, w(\{1\})) = 1$$

and, quite analogously, $\varepsilon(\{2\}) = 1$.

Regarding the coalition $I = \{1, 2\}$, there holds

$$\nu_{\succsim}(w(\{I\}), x_1 + x_2) > 0 \quad \text{for } x_1 + x_2 < 10$$

but then either $x_1 < 5$ and then $\nu_{\succsim}(x_1, w(\{1\})) = 0$ or $x_2 < 5$ and $\nu_{\succsim}(x_2, w(\{2\})) = 0$.

Consequently,

$$\varepsilon(I) = 0,$$

and the coalition I is not effective at all.

□

Example 6.6 Finally, let us consider the 3-players fuzzy coalition game from Example 6.3. Then, of course, $\varepsilon(\{i\}) = 1$ for $i = 1, 2, 3$, and also $\varepsilon(\{i, j\}) = 1$ for $i, j \in I$, $i \neq j$, as for, e.g., $x_i = 1$, $x_j = 1$, we have

$$\nu_{\succsim}(1, w(\{i\})) = 1, \quad \nu_{\succsim}(1, w(\{j\})) = 1,$$
$$\nu_{\succsim}(w(\{i, j\}), 2) = 1, \quad \nu_{\succsim}(2, w(\{i, j\})) = 1.$$

Due to the results of Example 6.3, $\varepsilon(I) = \gamma_C(I, w) = \frac{3}{5}$ and this possibility is attained for $\boldsymbol{x} = \left(\frac{4}{5}, \frac{4}{5}, \frac{4}{5}\right)$.

Let us verify this conclusion. Evidently, for this imputation $\boldsymbol{x}$, $x_1 + x_2 + x_3 = \frac{12}{5}$, $x_i + x_j = \frac{8}{5}$, $i, j \in I$. Then, using (2.11),(2.12), we have

$$\nu_{\succsim}\left(w(\{I\}), \frac{12}{5}\right) = 1, \quad \nu_{\succsim}\left(\frac{12}{5}, w(I)\right) = 1,$$
$$\nu_{\succsim}\left(\frac{8}{5}, w(\{i, j\})\right) = \frac{3}{5}, \quad \nu_{\succsim}\left(\frac{4}{5}, w(\{i\})\right) = 1$$

for $i, j \in I$, $i \neq j$. This means that $\varepsilon(I) \geq \frac{3}{5}$.

On the other hand, if $\boldsymbol{y} \in R^3$ is such that $y_1 + y_2 + y_3 > \frac{12}{5}$, then evidently $\nu_{\succsim}\left(w(\{I\}), \frac{12}{5}\right) < \frac{3}{5}$, and if $\boldsymbol{z} = (z_1, z_2, z_3) \in R^3$ fulfills $z_1 + z_2 + z_3 \leq \frac{12}{5}$ and $\boldsymbol{z} \neq \boldsymbol{x}$, then there exists at least one 2-players coalition $\{i, j\}$ for which $z_i + z_j < \frac{8}{5}$ and then $\nu_{\succsim}\left(\frac{8}{5}, w(\{i, j\})\right) < \frac{3}{5}$. It means that $\varepsilon(I)$ cannot be greater than $\frac{3}{5}$ and, consequently, $\varepsilon(I) = \frac{3}{5}$.

□

7. Fuzzy Convexity

The convexity in a deterministic coalition game with side-payments is recalled in Section 3.1 through (3.7). It is evident that there exist two equivalent definitional inequalities indicating the convexity, namely

$$v(K \cup L) + v(K \cap L) \geq v(K) + v(L)$$

and

$$v(K \cup L) \geq v(K) + v(L) - v(K \cap L)$$

for any $K, L \subset I$.

If we consider a fuzzy extension (I, w) of (I, v) with afuzzy characteristic function w and fuzzy quantities $w(K), w(L), w(K \cup L), w(K \cap L)$, then these inequalities turn into fuzzy relations whose possibilities need not be generally equivalent. Consequently, in the following paragraphs we choose the first one of them to be transformed into a fuzzy coalition game model.

7.1 General Properties

The convexity of a fuzzy extension of a coalition game with side-payments can be defined, analogously to [32], in the following way based on relation (3.7).

If (I, w) is a fuzzy extension of (I, v), then we say that (I, w) is *simply fuzzy convex* iff for every pair of coalitions $K, L \subset I$, there holds

$$w(K \cup L) \oplus w(K \cap L) \succsim w(K) \oplus w(L). \tag{7.1}$$

This relation is valid, due to (2.9), with the possibility $\pi(K, L)$ given by

$$\begin{aligned} & \nu_{\succsim}\left((w(K \cup L) \oplus w(K \cap L)), (w(K) \oplus w(L))\right) = \pi(K, L) = \\ = \; & \sup\left[\min\left(\mu_{(K \cup L)+(K \cap L)}(x), \mu_{K+L}(y)\right) : x, y \in R, x \geq y\right], \end{aligned} \tag{7.2}$$

where, analogously to (5.3) and (2.3),

$$\mu_{(K\cup L)+(K\cap L)}(x) = \sup_{y\in R}\left(\min(\mu_{K\cup L}(y), \mu_{K\cap L}(x-y))\right), \qquad (7.3)$$
$$\mu_{K+L}(x) = \sup_{y\in R}\left(\min(\mu_K(y), \mu_L(x-y))\right). \qquad (7.4)$$

The possibility that (I, w) is simply fuzzy convex is

$$\pi^0(I, w) = \min(\pi(K, L) : K, L \subset I). \qquad (7.5)$$

It is evident that any weakly fuzzy convex game is also weakly fuzzy superadditive with the possibility

$$\nu^0_{\text{super}}(I, w) \geq \pi^0(I, w)$$

as for disjoint coalitions $K, L \subset I$, $K \cap L = \emptyset$, $\mu_{K\cap L}(x) = \mu_\emptyset(x)$ for any $x \in R$ and $w(\emptyset) = \langle 0 \rangle$. It means that, for disjoint K, L, $\pi(K, L) = \overline{\nu}(K, L)$, and

$$\begin{aligned}\nu^0_{\text{super}}(I, w) &= \min(\overline{\nu}(K, L) : K, L \subset I, K \cap L = \emptyset) \geq \\ &\geq \min(\pi(K, L) : K, L \subset I) = \pi^0(I, w).\end{aligned}$$

The simple fuzzy convexity of (I, w) fulfills some properties mostly regarding its relation to the deterministic convexity of (I, v) and analogous to the ones shown for the superadditivity. It is, for example, evident, that $\pi(K, L) = \pi(L, K)$ for any $K, L \subset I$. Moreover, if (I, w) is a fuzzy extension of (I, v) and if (I, v) is convex, then (I, w) is fuzzy convex with possibility $\pi^0(I, w) = 1$.

If (I, w) and (I, w') are fuzzy extensions of a game (I, v) with membership functions μ_K and μ'_K for $w(K)$ and $w'(K)$, respectively, and if $\mu_K(x) \geq \mu'_K(x)$ for all $K \subset I$ and $x \in R$, then $\pi^0(I, w) \geq \pi^0(I, w')$.

For every game (I, v) there exists its fuzzy extension (I, w) such that $\pi^0(I, w) = 1$. But, on the other hand, if (I, w) is a fuzzy extension of (I, v) such that for every coalition $K \subset I$, $\mu_K(x) = 1$ iff $x = v(K)$, then $\pi^0(I, w) = 1$ iff (I, w) is convex as shown in [32].

In the introductory part of this section we have mentioned that the fuzzification of the concept of convexity based on the inequalities

$$v(K \cup L) \geq v(K) + v(L) - v(K \cap L)$$

is generally different from the one based on

$$v(K \cup L) + v(K \cap L) \geq v(K) + v(L)$$

and used above. This difference is based on the fact that the possibilities of fuzzy relations

$$w(k \cup L) \oplus w(K \cap L) \succsim w(K) \oplus w(L)$$

and

$$w(K \cup L) \succsim w(K) \oplus w(L) \oplus (-w(K \cap L)) \tag{7.6}$$

(where (2.4) can be noticed) are not equivalent as follows from (2.9).

If we denote for $K, L \subset I$

$$\pi^*(K,L) = \nu_{\succsim}(w(K \cup L), (w(K) \oplus w(L) \oplus (-w(K \cap L))))$$

and $\pi^*(I,w) = \min(\pi^*(K,L) : K, L \subset I)$, then, generally, $\pi^*(K,L) \neq \pi(K,L)$ and $\pi^*(I,w) \neq \pi(I,w)$ but all the statements introduced in this subsection for $\pi(K,L)$ and $\pi^0(I,w)$ are valid also for $\pi^*(K,L)$ and $\pi^*(I,w)$.

Anyway, the equality between $\pi^0(I,w)$ and $\pi^*(I,w)$ is possible for many cases of fuzzy coalition games (e.g., for games with triangular membership functions) as shown in the following example.

Example 7.1 Let us consider, for illustration, the 3-players fuzzy coalition game (I,w), already dealt with in Example 6.3 and Example 6.6 which is a fuzzy extension of the deterministic game (I,v) shown in Example 3.2. It means that $I = \{1,2,3\}$, $v(\{i\}) = 0$, $v(\{i,j\}) = 2$, $i,j \in I$, $i \neq j$, $v(I) = 2$.

This game is not convex as shown in Example 6.3. The membership functions of its fuzzy extension are for $i,j \in I$, $i \neq j$,

$$\begin{aligned}
\mu_{\{i\}}(x) &= x+1 && \text{for} && x \in [-1,0], \\
&= 1 && \text{for} && x \in [0,1], \\
&= 0 && \text{otherwise,} &&
\end{aligned}$$

$$\begin{aligned}
\mu_{\{i,j\}}(x) = \mu_I(x) &= x-1 && \text{for} && x \in [1,2], \\
&= 3-x && \text{for} && x \in [2,3], \\
&= 0 && \text{otherwise.} &&
\end{aligned}$$

Let us verify both the above concepts of its convexity. It is not difficult to compute that

$$\begin{aligned}
\pi(K,L) &= \frac{1}{2} \text{ for the pairs of coalitions} \\
&\quad (\{1,2\},\{1,3\}), (\{1,2\},\{2,3\}), (\{1,3\},\{2,3\})
\end{aligned}$$

and $\pi(K,L) = 1$ for other pairs of coalitions. It means that

$$\pi^0(I,w) = \frac{1}{2}.$$

In this case also $\pi^*(K,L) = \pi(K,L)$ for all coalitions $K, L \subset I$ and, consequently,

$$\pi^*(I,w) = \frac{1}{2}.$$

□

The above equality between $\pi^0(I,w)$ and $\pi^*(I,w)$ is not random. It follows from the triangularity of membership function μ_K for $K \subset I$, $K \neq \emptyset$, and it is fulfilled for any such a fuzzy coalition game.

It is quite interesting to compute the possibilities $\pi^0(I,w)$ and, of course, also $\pi^*(I,w)$ for different simple modifications of the previous game. The development of these values depending on the variability of $v(\{i\})$ and $w(\{i\})$ is at least partly shown by the following example.

Example 7.2 Let us consider a set of 3-players coalition games (I, v_m), $m = 1, \ldots, 15$ in which $v_m(I) = 2$, $v_m(\{i,j\}) = 2$ for $i,j \in I$, $i \neq j$, $m = 1, \ldots, 15$, and for the 1-player coalitions $\{i\}$, $i = 1,2,3$, there is

$$\begin{array}{lllll} v_1(\{i\}) = -2, & v_2(\{i\}) = -1, & v_3(\{i\}) = -\frac{1}{2}, & v_4(\{i\}) = 0, & v_5(\{i\}) = \frac{1}{5}, \\ v_6(\{i\}) = \frac{1}{2}, & v_7(\{i\}) = \frac{3}{5}, & v_8(\{i\}) = \frac{3}{4}, & v_9(\{i\}) = \frac{4}{5}, & v_{10}(\{i\}) = 1, \\ v_{11}(\{i\}) = \frac{6}{5}, & v_{12}(\{i\}) = \frac{3}{2}, & v_{13}(\{i\}) = \frac{8}{5}, & v_{14}(\{i\}) = 2, & v_{15}(\{i\}) = 3. \end{array}$$

Also the fuzzy extensions of the games (I, v_m) only slightly modify the pattern used in Example 7.1. For 2-players and 3-players coalitions the fuzzy quantities $w(\{i,j\})$ $(= w(I))$ and their membership functions are

$$\begin{array}{rclll} \mu_{\{i,j\}}(x) & = & \mu_I(x) = x - 1 & \text{for} & x \in [1,2], \\ & = & 3 - x & \text{for} & x \in [2,3], \\ & = & & \text{otherwise.} & \end{array}$$

The membership functions $\mu^{(m)}_{\{i\}}$ of $w_m(\{i\})$, $m = 1, 2, \ldots, 15$, are generally

$$\begin{array}{rclll} \mu^{(m)}_{\{i\}}(x) & = & x + 1 - v_m(\{i\}) & \text{for} & x \in [v_m(\{i\}) - 1, v_m(\{i\})], \\ & = & -x + 1 - v_m(\{i\}) & \text{for} & x \in [v_m(\{i\}), v_m(\{i\}) + 1], \\ & = & 0 & \text{otherwise.} & \end{array}$$

Then the possibilities of fuzzy convexity of the games (i, w_m), $m = 1, \ldots, 15$, are summarized in Table 7.2.

It is worth mentioning that the values which are significant for the computation of $\pi^0(I, w_m)$ and $\pi^*(I, w_m)$ are the possibilities of the inequalities:

Table 7.1. Values of $\pi(K, L)$ and $\pi^0(I, w_m)$, $m = 1, \ldots, 15$ in Example 7.2, $i, j, k \in I$, $i \neq j \neq k \neq i$. Possibilities of fuzzy convexity of the games (i, w_m), $m = 1, \ldots, 15$

m	$v_m(\{i\})$	$\pi_m(\{i\},\{j\})$	$\pi_m(\{i\},\{j,k\})$	$\pi_m(\{i,j\},\{j,k\})$	$\pi^0(I, w_m)$
1	–2	1	1	0	$0 = 0$
2	–1	1	1	1/4	$1/4 = 0,25$
3	–1/2	1	1	3/8	$3/8 = 0,375$
4	0	1	1	1/2	$1/2 = 0,5$
5	1/5	1	14/15	11/20	$11/20 = 0,55$
6	1/2	1	5/6	5/8	$5/8 = 0,625$
7	3/5	1	4/5	13/20	$13/20 = 0,65$
8	3/4	1	3/4	11/16	$11/16 = 0,6875$
9	4/5	1	11/15	7/10	$7/10 = 0,7$
10	1	1	2/3	3/4	$2/3 = 0,6667$
11	6/5	13/15	3/5	4/5	$3/5 = 0,6$
12	3/2	2/3	1/2	7/8	$1/2 = 0,5$
13	8/5	3/5	7/15	9/10	$7/15 = 0,4667$
14	2	1/3	1/3	1	$1/3 = 0,333$
15	3	0	0	1	0

$$w_m(\{i,j\}) \succsim w_m(\{i\}) \oplus w_m(\{j\})$$
$$\text{denoted } \pi(\{i\},\{j\}) = \pi^*(\{i\},\{j\}),$$

$$w_m(I) \succsim w_m(\{i\}) \oplus w_m(\{j,k\})$$
$$\text{denoted } \pi(\{i\},\{j,k\}) = \pi^*(\{i\},\{j,k\}),$$

$$w_m(I) \oplus w_m(\{j\}) \succsim w_m(\{i,j\}) \oplus w_m(\{j,k\})$$
$$\text{denoted } \pi(\{i,j\},\{j,k\}),$$

$$w_m(I) \succsim w_m(\{i,j\}) \oplus w_m(\{j,k\}) \oplus (-w_m(\{j\}))$$
$$\text{denoted } \pi^*(\{i,j\},\{j,k\}).$$

The other values $\pi(K,L)$ and $\pi^*(K,L)$ are identically equal to 1. Let us also mention that the games (I, v_4) and (I, w_4) are those investigated in Example 7.1.

Remember that, due to the triangular shape of μ_K, for all $K, L \subset I$, $\pi(K,L) = \pi^*(K,L)$.

□

In the case of superadditivity and subadditivity, we have seen that while in the deterministic games the validity of definitional inequalities

$$v(K \cup L) \geq v(K) + v(L) \quad \text{and} \quad w(K \cup L) \leq v(K) + v(L)$$

for disjoint pairs $K, L \subset I$ implies analogous relations for larger classes of disjoint coalitions, in the fuzzy coalition games it need not be so. A similar situation appears in the case of convexity even if in this case the general relation is much more structured.

In the following paragraphs we suggest a generalization of the fuzzy convexity concept which better corresponds to the structure of values of fuzzy characteristic functions. Namely, we suggest the fuzzy convexity derived from the relations between at least (including more than) two coalitions. We focus our attention on the generalization of (7.1). It is evident that a very similar procedure can be used to generalize (7.6).

This modification does not bring any qualitatively new results and we will not be interested in it. Its main significance consists in its sufficiency for the existence of a non-empty core. As we can see in the following Section 7.2, the weak fuzzy convexity completely satisfies this purpose and, consequently, the following suggestion of a generalized convexity satisfies the demands of formal completeness of the presented fuzzy coalition games theory.

If (I, w) is a fuzzy extension of (I, v), then we say that it is *fuzzy convex* iff for any set of coalitions there holds $\{K_1, K_2, \ldots, K_m\} = \mathcal{L}$ the relation

$$w(K_1 \cup \cdots \cup K_m) \oplus \left(\sum\nolimits_{\mathcal{J} \subset \mathcal{L}}^{\oplus} w\left(\bigcap\nolimits_{K_j \in \mathcal{J}} K_j\right)\right) \succsim w(K_1) \oplus \cdots \oplus w(K_m) \tag{7.7}$$

where $\sum^{\oplus}$ denotes the summation on the set of fuzzy quantities in which the addition operation $\oplus$ is used.

The possibility of this relation can be computed using (2.9) and it will be denoted by $\pi(K_1, \ldots, K_m)$.

The possibility that (I, w) is fuzzy convex is denoted by

$$\pi(I, w) = \min\left(\pi(K_1, \ldots, K_m) : m = 2, 3, \ldots,\ K_1, \ldots, K_m \subset I\right). \tag{7.8}$$

It is evident that the value $\pi(K_1, K_2)$ derived using (7.7) is identical with $\pi(K_1, K_2)$ derived by (7.2) and, consequently,

$$\pi(I, w) \leq \pi^0(I, w).$$

The computation of the value $\pi(I, w)$ for a fuzzy coalition game leads to a methodological problem. The classes of coalitions (which need not be disjoint) can be arbitrarily extended. In fact, a repetition of one coalition does not bring any new information. On the contrary, a repeated calculation with one membership function only increases the extent of uncertainty (support of the membership function) of the result and, due to the partwise monotonicity of the membership functions, the values of the membership function of the result. This means that $\pi(K_1, \ldots, K_m) \geq \pi(K_1, \ldots, K_{\ell-1}, K_{\ell+1}, \ldots, K_m)$ if $K_\ell = K_k$. for some $k \in \{1, \ldots, m\}$, $k \neq \ell$.

Even if this principle is respected, the values of $\pi(I, w)$ usually do not significantly differ from $\pi^0(I, w)$.

Example 7.3 To illustrate at least partly the behavior of the values $\pi(K_1, \ldots, K_m)$ and their influence on the value $\pi(I, w)$, we process three of the games already investigated in the previous Example 7.2, namely (I, v_2), (I, v_4) and (I, v_{10}), and their fuzzy extensions (I, w_2), (I, w_4) and (I, w_{10}). We already know the values of $\pi(K, L)$ for any pair of coalitions K, L. As the complexity of the procedure enormously grows for larger classes of coalitions, we limit our attention to their triples K, L, M and possibilities $\pi(K, L, M)$. The symmetry of the considered games allows us to simplify the list of the considered classes of coalitions to

$$(\{1\}, \{2\}, \{3\}), (\{i\}, \{j\}, \{i, j\}), (\{i\}, \{j, k\}, \{i, k\}),$$
$$(\{i\}, \{i, j\}, \{i, k\}), (\{1, 2\}, \{1, 3\}, \{2, 3\}), (\{i\}, \{j\}, I),$$
$$(\{i\}, \{i, j\}, I), (\{i\}, \{j, k\}, I), (\{i, j\}, \{j, k\}, I).$$

It is not so difficult to verify that

$$\pi_2(K,L,M) = \pi_4(K,L,M) = \pi_{10}(K,L,M) = 1$$

for the triples K, L, M being

$$(\{i\},\{j,k\},\{j\}), (\{i\},\{j\},\{i,j\}),$$
$$(\{i\},\{j,k\},\{i,k\}), (\{i\},\{j\},I), (\{i\},\{j,k\},I).$$

The other values are more significant. Namely, for (I, w_2)

$$\pi_2(\{1,2\},\{2,3\},\{1,3\}) = 0, \quad \pi_2(\{i\},\{i,j\},I) = 6/7,$$
$$\pi_2(\{i,j\},\{j,k\},I) = 6/7, \quad \pi_2(\{i\},\{i,j\},\{i,k\}) = 8/9,$$
$$\pi_2(\{1\},\{2\},\{3\}) = 1.$$

The minimum of these values is

$$0 < \pi^0(I, w_2) = 1/4,$$

and the triples of coalitions influence the possibility of convexity.

What concerns the game (I, w_{10}), we have

$$\pi_{10}(\{1\},\{2\},\{3\}) = 3/4, \quad \pi_{10}(\{1,2\},\{1,3\},\{2,3\}) = 6/7,$$
$$\pi_{10}(\{i\},\{i,j\},I) = \pi_{10}(\{i,j\},\{j,k\},I) = \pi_{10}(\{i\},\{i,j\},\{i,k\}) = 1.$$

The minimum of these values is

$$3/4 > \pi^0(I, w_{10}) = 2/3$$

and the triples of coalitions do not influence the possibility of convexity.

Finally, for the game (I, w_4), we have

$$\begin{aligned}\pi_4(\{1\},\{2\},\{3\}) &= \pi_4(\{i\},\{i,j\},I) = \pi_4(\{i,j\},\{j,k\},I) = \\ &= \pi_4(\{i\},\{i,j\},\{i,k\}) = 1, \\ \pi_4(\{1,2\},\{1,3\},\{2,3\}) &= 3/7.\end{aligned}$$

In this case the minimum of these values is

$$3/7 < \pi^0(I, w_4) = 1/2,$$

and the possibility $\pi_4(\{1,2\},\{1,3\},\{2,3\})$ influences the value $\pi(I, w_4)$.

□

Analogously to the previous concept of a weak fuzzy convexity, we can also easily derive that, for any (I, w), there holds

$$\pi(I,w) \leq \nu_{\text{super}}(I,w).$$

Other properties introduced for $\pi^0(I,w)$ remain valid for $\pi(I,w)$ as well. Namely, if (I,v) is convex in the deterministic case, then $\pi(I,w) = 1$. For any (I,v) there exists a fuzzy extension (I,w) such that $\pi(I,w) = 1$. On the other hand, if for any $K \subset I$, $\mu_K(x) = 1$ iff $x = v(K)$, then $\pi(I,w) = 1$ iff (I,v) is convex.

7.2 Relations to the Fuzzy Core

As mentioned in Subsection 3.1, the convexity of a deterministic coalition game with side-payments implies the existence of a non-empty core. If (I,w) is a fuzzy extension of a convex deterministic game, then $\pi^0(I,w) = \pi^*(I,w) = \pi(I,w) = 1$. Moreover, the core C of (I,v) is non-empty and it is realized by the coalition structure formed by the single all-players coalition. Hence, I is effective. It means that also the fuzzy core C_F of (I,w) is non-empty, $\gamma_C(I,w) = 1$ and also $\varepsilon(I) = 1$.

On the level of certain phenomena the relation between the (fuzzy) convexity and (fuzzy) core is completely analogous to the one being valid for the deterministic case. It is natural to ask whether some analogous implication is valid also for a more uncertain situation. If, for example, $\varepsilon(I) \geq \pi^0(I,w)$ or $\varepsilon(I) \geq \pi^*(I,w)$ or $\varepsilon(I) \geq \pi(I,w)$ where $\varepsilon(I)$, the possibility that I is effective means also the possibility that the core of (I,w) is non-empty and realized within the coalition I.

An intuitive expectation supports the validity of such inequalities. The following consideration is not a proof, only an intuitive argument, but it seems to be quite convincing.

If (I,w) is fuzzy convex with possibility $\pi(I,w)$, then there exists for every coalition K a real number $v_K^* \in R$ such that

$$\mu_K(v_K^*) \geq \pi(I,w), \quad \min_{K \subset I} \mu_K(v_K^*) = \pi(I,w).$$

This means that with the possibility $\pi(I,w)$ there exists a convex deterministic game (I,v^*) with $v^*(K) = v_K^*$, for all $K \subset I$. This game, being convex, has a non-empty deterministic core. It means that, at least with possibility $\pi(I,w)$ (or a greater one), the fuzzy quantities $w(K)$ attain values v_K^* which fulfil the properties of the core that is realized by the all-players coalition I. It means that $\varepsilon(I)$ is to be at least equal to $\pi(I,w)$.

The examples which were presented in the previous subsection can at least partly illustrate the validity of the above consideratioins. Note, however, that they, in fact, do not offer the values $\pi(I,w)$ as the procedure used there does not include the computation of $\pi(K_1,\ldots,K_m)$, for all classes $K_1,\ldots,K_m$ of different coalitions. However, it is quite interesting to compare the results derived in Examples 7.2 and 7.3 with our previous argument.

Table 7.2. Possibilities of core, effectivity and convexity in Example 7.4. Values of possibilities $\pi^0(I, w_m)$, $\varepsilon_m(I)$ and $\gamma_C(I, w_m)$

m	$v_m(\{i\})$	$\pi^0(I, w_m)$	$\varepsilon_m(I)$	$\gamma_C(I, w_m)$	vector realizing $\varepsilon_m(I)$
1	–2	0	3/5	3/5	(4/5,4/5,4/5)
2	–1	1/4	3/5	3/5	(4/5,4/5,4/5)
3	–1/2	3/8	3/5	3/5	(4/5,4/5,4/5)
4	0	1/2	3/5	3/5	(4/5,4/5,4/5)
5	1/5	11/20	3/5	3/5	(4/5,4/5,4/5)
6	1/2	5/8	3/5	3/5	(4/5,4/5,4/5)
7	3/5	13/20	3/5	3/5	(4/5,4/5,4/5)
8	3/4	11/16	3/5	3/5	(4/5,4/5,4/5)
9	4/5	7/10	3/5	3/5	(4/5,4/5,4/5)
10	1	2/3	3/5	1	(4/5,4/5,4/5)
11	6/5	3/5	3/5	1	(4/5,4/5,4/5)
12	3/2	1/2	3/8	1	(7/8,7/8,7/8)
13	8/5	7/15	3/10	1	(9/10,9/10,9/10)
14	2	1/3	0	1	—
15	3	0	0	1	—

Example 7.4 Let us consider, again, the fifteen coalition games (I, v_m) and their fuzzy extensions (I, w_m), $m = 1, \dots, 15$, which were investigated in Example 7.2. It is not difficult to verify, analogously to the procedure used in Example 6.3 and Example 6.6, that the values of possibilities $\pi^0(I, w_m)$, $\varepsilon_m(I)$ and $\gamma_C(I, w_m)$ are those ones introduced in Table 7.4.

Evidently, the games (I, w_1), (I, w_2), (I, w_3), (I, w_4), (I, w_5) and (I, w_{11}), fulfil our idea about

$$\varepsilon_m(I) \geq \pi^0(I, w_m)$$

while the other games do not.

The results derived in Example 7.3 show that this conclusion is not corrected even if we include the triples of coalitions into the procedure.

□

A deeper analysis of these examples is rather ambiguous. Our intuitive conclusions were not confirmed at least at the level of 2-coalitions (and in some cases 3-coalitions) classes. It is possible that the roots of that can be found in purely formal procedures.

If we analyze the effectivity (and core), we deal with fuzzy quantities and their orderings with crisp numbers without any transformation of their membership functions. On the other hand, in the computation of the convexity possibilities we compare sums of fuzzy quantities. The application of formula (2.3) essentially increases the extent of supports of the membership functions and, as a consequence of their partwise monotonicity, also their values. This leads to an increase of the values $\pi(K, L), \pi^*(K, L)$ and (even more effectively) of $\pi(K_1, \ldots, K_m)$. Maybe, using another procedure for the calculation of the sums of fuzzy quantities, the results could better correspond to our intuition.

Such an alternative approach to the operations over fuzzy quantities is briefly mentioned in Chapter 16 and it is at least worth a more detailed analysis (see also Example 3.11).

8. Fuzzy Balancedness

If we remember Section 3.1, namely the two relations (3.8) and (3.9), we see that the balancedness of a set of coalitions $\mathcal{L} = \{L_1, \ldots, L_m\}$ is a fully deterministic phenomenon. The fuzzified characteristic function w of (I, w) can only influence the formal properties of the formula that are analogous to (3.9).

Let (I, w) be a fuzzy extension of (I, v). We say that (I, w) is *fuzzy* iff for every balanced set of coalitions $\mathcal{L} = \{L_1, \ldots, L_m\}$ with coefficients $\alpha_1, \ldots, \alpha_m$ fulfilling (3.8) the relation

$$w(I) \precsim \alpha_1 \cdot w(L_1) \oplus \alpha_2 \cdot w(L_2) \oplus \cdots \oplus \alpha_m \cdot w(L_m) \tag{8.1}$$

holds.

The possibility of (8.1) is, in accordance with (2.9), equal to

$$\begin{aligned} \beta(\mathcal{L}, \alpha) &= \nu_{\succsim}(w(I), \alpha_1 \cdot w(L_1) \oplus \cdots \oplus \alpha_m \cdot w(L_m)) = \\ &= \sup_{\substack{x,y \in R \\ x \geq y}} \left[\min(\mu_I(x), \mu_{\mathcal{L},\alpha}(y))\right], \end{aligned} \tag{8.2}$$

where $\mathcal{L} = \{L_1, \ldots, L_m\} \subset 2^I$, $\alpha = (\alpha_1, \ldots, \alpha_m) \in R^m$ are the balancedness coefficients of $\mathcal{L}$, and $\mu_{\mathcal{L},\alpha}$ is the membership function of the fuzzy quantity $\alpha_1 \cdot w(L_1) \oplus \cdots \oplus \alpha_m \cdot w(L_m)$. (As the operations over the set $\mathbb{R}$ are associative and distributive, cf. [25], the notation used above is correct.)

Hence, the possibility that (I, w) is fuzzy balanced is equal to the possibility

$$\begin{aligned} \beta(I, w) = \; & \min(\beta(\mathcal{L}, \alpha) : \\ & \mathcal{L} \subset R^I, \mathcal{L} \text{ is balanced with} \\ & \text{the vector of coefficients } \alpha). \end{aligned}$$

Properties of the fuzzy balancedness remind analogous properties which we already know from the sections on superadditivity or convexity.

8.1 General Properties

It is not difficult to derive relations between the deterministic balancedness of the game (I, v) and the fuzzy balancedness of its fuzzy extension (I, w).

If, for any coalition $K \subset I$, $\mu_K(x) = 1$ iff $x = v(K)$, then for any balanced set of coalitions $\mathcal{L} = \{L_1, \ldots, L_m\}$ with the vector of coefficients $\alpha = (\alpha_1, \ldots, \alpha_m)$ the possibility $\beta(\mathcal{L}, \alpha) = 1$ if and only if

$$v(I) \geq \alpha_1 \cdot v(L_1) + \cdots + \alpha_m \cdot v(L_m).$$

This means that for (I, w) fulfilling the above condition of unimodality, $\beta(I, w) = 1$ iff (I, v) is balanced.

Obviously, every partition of I into disjoint coalitions is a balanced set of coalitions with coefficients $\alpha_j = 1$. It means, due to (8.3) and (5.4), that for every coalition structure $\mathcal{K} = (K_1, \ldots, K_m)$ we have

$$\overline{\nu}(K_1, \ldots, K_m) \geq \beta(I, w).$$

It is also easy to verify that if (I, w) and (I, w') are fuzzy extensions of (I, v) with the membership functions μ_K, μ'_K, $K \subset I$, and such that for any $K \subset I$, $\mu_K(x) \geq \mu'_K(x)$, for all $x \in R$, then for any balanced set of coalitions $\mathcal{L}$ general properties of the ordering relation $\succsim$ imply $\beta(\mathcal{L}, \alpha) \geq \beta'(\mathcal{L}, \alpha)$ where

$$\begin{aligned} \beta(\mathcal{L}, \alpha) &= \nu_{\succsim}\left(w(I), \alpha_1 \cdot w(L_1) \oplus \cdots \oplus \alpha_m \cdot w(L_m)\right), \\ \beta'(\mathcal{L}, \alpha) &= \nu_{\succsim}\left(w'(I), \alpha_1 \cdot w'(L_1) \oplus \cdots \oplus \alpha_m \cdot w'(L_m)\right). \end{aligned}$$

Consequently, $\beta(I, w) \geq \beta(I, w')$.

Also in the case of balancedness for every game (I, v) there exists its fuzzy extension (I, w) such that $\beta(I, w) = 1$. This fuzzy extension can be obtained if the sets

$$\{x \in R : \mu_K(x) = 1\}$$

are large enough for all coalitions $K \subset I$.

By investigating the fuzzy balancedness we encounter a problem modification that already occured in the case of superadditivity. Namely, the increase of the extent of the balanced classes of coalitions (e.g., the union of balanced classes of coalitions is balanced as well) influences the shape of the membership function of the weighted sum on the right hand side of (8.1), while in the deterministic games such a union does not influence the balancedness of the game. More precisely, if $\mathcal{L}_1 = (L_1, \ldots, L_m)$ is balanced with the vector of coefficients $\boldsymbol{\alpha}^{(1)} = (\alpha_1^{(1)}, \ldots, \alpha_m^{(1)})$ and $\mathcal{L}_2 = (K_1, \ldots, K_n)$ is balanced with the vector $\boldsymbol{\alpha}^{(2)} = (\alpha_1^{(2)}, \ldots, \alpha_n^{(2)})$, then $\mathcal{L} = \mathcal{L}_1 \cup \mathcal{L}_2$ is also balanced with the vector

$$\boldsymbol{\alpha} = \left(\alpha_1^{(1)}, \ldots, \alpha_m^{(1)}, \alpha_1^{(2)}, \ldots, \alpha_n^{(2)}\right) = (\alpha_1, \ldots, \alpha_{m+n})$$

where

$$\begin{aligned} \alpha_k &= \lambda \cdot \alpha_k^{(1)} && \text{for } k = 1, \ldots, m, \\ \alpha_k &= (1-\lambda)\, \alpha_k^{(2)} && \text{for } k = m+1, \ldots, m+n,\ \lambda \in (0,1). \end{aligned}$$

If we denote by $\beta(\mathcal{L}_1, \boldsymbol{\alpha}^{(1)}), \beta(\mathcal{L}_2, \boldsymbol{\alpha}^{(2)})$ and $\beta(\mathcal{L}, \boldsymbol{\alpha})$ the possibilities (8.2) for the respective balanced classes, then it is not difficult to derive that

$$\beta(\mathcal{L}, \boldsymbol{\alpha}) \geq \min\left(\beta(\mathcal{L}_1, \boldsymbol{\alpha}^{(1)}), \beta(\mathcal{L}_2, \boldsymbol{\alpha}^{(2)})\right).$$

Consequently, the enlarging of the balanced classes of coalitions by means of their union does not influence the possibility $\beta(I, w)$ of the balancedness of the game (I, w), and we can limit our procedures to the simple balanced classes of coalitions which cannot be reduced. Let us illustrate the concepts introduced above by several simple examples.

Example 8.1 Consider the coalition game (I, v) already investigated in Example 6.1, i.e. $I = (\{1,2\})$, $v(\{1\}) = 1$, $v(\{2\}) = 2$, $v(I) = 5$, and its fuzzy extension (I, w) introduced in Example 6.1, and

$$\begin{aligned} \mu_{\{1\}}(x) &= x && \text{for} && x \in [0,1], \\ &= 2 - x && \text{for} && x \in [1,2], \\ &= 0 && \text{otherwise}, \end{aligned}$$

$$\begin{aligned} \mu_{\{2\}}(x) &= x - 1 && \text{for} && x \in [1,2], \\ &= 3 - x && \text{for} && x \in [2,3], \\ &= 0 && \text{otherwise}, \end{aligned}$$

$$\begin{aligned} \mu_{I}(x) &= x - 4 && \text{for} && x \in [4,5], \\ &= 6 - x && \text{for} && x \in [5,6], \\ &= 0 && \text{otherwise}. \end{aligned}$$

If we consider only the classes of coalitions in which each coalition is at most once, then there exist the following balanced classes:

$(\{1\}, \{2\})$	–	with the vector of weights $\boldsymbol{\alpha} = (1, 1)$,
(I)	–	with the 1-element vector $\boldsymbol{\alpha} = (1)$,
$(I, \{1\}, \{2\})$	–	with any vector of weights $\boldsymbol{\alpha} = (\lambda, 1-\lambda, 1-\lambda)$ for $\lambda \in [0, 1]$.

The last class is the union of the previous ones and, due to the former comment, we need not be interested in it. Anyway, for illustration, we show also its possibility $\beta(\mathcal{L}, \boldsymbol{\alpha})$. As it can be easily verified, these possibilities are

$$\beta((\{1\}, \{2\}), (1,1)) = 1,\ \beta((I), (1)) = 1,\ \beta((I, \{1\}, \{2\})\,(\lambda, 1-\lambda, 1-\lambda)) = 1.$$

It means that $\beta(I, w) = 1$. It is also easy to verify that (I, v) is balanced in the deterministic sense.

□

Example 8.2 In this example we are interested in the coalition game (I,v) and its fuzzy extension (I,w) already investigated in Example 6.2. Let us remember that it is a 2-players game with $v(\{i\}) = 6$, $i = 1,2$, $v(I) = 9$, and for $i = 1,2$

$$\begin{aligned} \mu_{\{i\}}(x) &= x-5 \quad \text{for} \quad x \in [5,6], \\ &= 7-x \quad \text{for} \quad x \in [6,7], \\ &= 0 \quad \text{otherwise}, \end{aligned}$$

$$\begin{aligned} \mu_I(x) &= x-8 \quad \text{for} \quad x \in [8,9], \\ &= 10-x \quad \text{for} \quad x \in [9,10], \\ &= 0 \quad \text{otherwise}. \end{aligned}$$

Then, it is easy to verify that

$$\beta(\{1\},\{2\},(1,1)) = 0, \quad \beta((I),(1)) = 1$$

and, e.g., for $\lambda = \frac{1}{2}$

$$\beta\left((I,\{1\},\{2\}),\left(\frac{1}{2},\frac{1}{2},\frac{1}{2}\right)\right) = 5/8.$$

But, $\beta(I,w) = 0$.

□

The simple procedure used for the 2-players games is not much more complicated than the one for the case of 3-players games.

Example 8.3 Let us consider a 3-players coalition game (I,v) and its fuzzy extension (I,w) already investigated in Example 6.3. It means that $v(\{i\}) = 0$, $v(\{i,j\}) = 2$, $i,j \in I$, $i \neq j$, $v(I) = 2$, and

$$\begin{aligned} \mu_{\{i\}}(x) &= x+1 \quad \text{for} \quad x \in [-1,0], \\ &= 1-x \quad \text{for} \quad x \in [0,1], \\ &= 0 \quad \text{otherwise}, \end{aligned}$$

$$\begin{aligned} \mu_{\{i,j\}}(x) = \mu_I(x) &= x-1 \quad \text{for} \quad x \in [1,2], \\ &= x-3 \quad \text{for} \quad x \in [2,3], \\ &= 0 \quad \text{otherwise}. \end{aligned}$$

The effective balanced classes of coalitions in this game are

$$\begin{array}{lll} (I) & \text{with} & \boldsymbol{\alpha} = (1) \\ (\{i\},\{j,k\}) & \text{with} & \boldsymbol{\alpha} = (1,1), \ i,j,k \in I, \ i \neq j \neq k \neq i, \\ (\{1\},\{2\},\{3\}) & \text{with} & \boldsymbol{\alpha} = (1,1,1) \\ (\{1,2\},\{1,3\},\{2,3\}) & \text{with} & \boldsymbol{\alpha} = \left(\frac{1}{2},\frac{1}{2},\frac{1}{2}\right). \end{array}$$

The other balanced classes in which every coalition exists at most once, namely

$$(I, \{1\}, \{2\}, \{3\}),\ (I, \{i\}, \{j, k\}),\ (I, \{1, 2\}, \{1, 3\}, \{2, 3\})$$
$$(\{i\}, \{j, k\}, \{i, j\}, \{j\}),\ (I, \{i\}, \{j\}, \{i, k\}, \{j, k\}),$$
$$(\{1\}, \{2\}, \{3\}, \{1, 2\}, \{1, 3\}, \{2, 3\}),\ (I, \{1\}, \{2\}, \{3\}, \{1, 2\}, \{1, 3\}, \{2, 3\})$$

are the unions of the previous effective four classes and they are not interesting.

It is easy to verify that

$$\beta((I), (1)) = 1, \quad \beta((\{1\}, \{2\}, \{3\}), (1, 1, 1)) = 1,$$
$$\beta((\{i\}, \{j, k\}), (1, 1)) = 1, \quad \beta\left((\{1, 2\}, \{1, 3\}, \{2, 3\}), \left(\tfrac{1}{2}, \tfrac{1}{2}, \tfrac{1}{2}\right)\right) = \tfrac{3}{4}$$

and, consequently, $\beta(I, w) = \frac{3}{4}$.

□

Example 8.4 Finally, we will deal with a 3-players coalition game (I, v) with $I = \{1, 2, 3\}$, $v(\{i\}) = 1$, $v(\{i, j\}) = 3$, $i, j \in I$, $i \neq j$, $v(I) = 5$. Evidently, its core C is non-empty and attained by the coalition I.

Let us consider a fuzzy extension (I, w) of (I, v) such that

$$\begin{aligned} \mu_{\{i\}}(x) &= x && \text{for} && x \in [0, 1], \\ &= 2 - x && \text{for} && x \in [1, 2], \\ &= 0 && \text{otherwise,} \end{aligned}$$

$$\begin{aligned} \mu_{\{i,j\}}(x) &= \tfrac{1}{2}x - \tfrac{1}{2} && \text{for} && x \in [1, 3], \\ &= \tfrac{5}{2} - \tfrac{1}{2}x && \text{for} && x \in [3, 5], \\ &= 0 && \text{otherwise,} \end{aligned}$$

$$\begin{aligned} \mu_I(x) &= \tfrac{1}{3}x - \tfrac{2}{3} && \text{for} && x \in [2, 5], \\ &= \tfrac{8}{3} - \tfrac{1}{3}x && \text{for} && x \in [5, 8], \\ &= 0 && \text{otherwise.} \end{aligned}$$

Then, it is easy to see that

$$\beta((I), (1)) = 1, \quad \beta((\{1\}, \{2\}, \{3\}), (1, 1, 1)) = 1,$$
$$\beta((\{i\}, \{j, k\}), (1, 1)) = 1, \quad \beta\left((\{1, 2\}, \{1, 3\}, \{2, 3\}), \left(\tfrac{1}{2}, \tfrac{1}{2}, \tfrac{1}{2}\right)\right) = 1$$

and, consequently, $\beta(I, w) = 1$.

□

Example 8.5 It is useful to compute the possibilities $\beta(\mathcal{L}, \boldsymbol{\alpha})$ and $\beta(I, w)$ for the set of 15 games already investigated in Example 7.2 (and also in Example 7.4). Let us recall that for all coalition games (I, v_m), $m = 1, \ldots, 15$, $v(\{i, j\}) = v(I) = 2$, $i, j \in I$, $i \neq j$, and the $v(\{i\})$'s attain the values summarized in Table 7.2.

The fuzzy extensions (I, w_m) of (I, v_m) are such that, for $w_m(\{i, j\})$ and $w_m(I)$,

$$
\begin{aligned}
\mu^{(m)}_{\{i,j\}}(x) = \mu^{(m)}_I(x) &= x-1 \quad \text{for} \quad x \in [1,2],\\
&= 3-x \quad \text{for} \quad x \in [2,3],\\
&= 0 \quad \text{otherwise},\\
&\qquad m = 1,\ldots,15,
\end{aligned}
$$

and for $w_m(\{i\})$, generally,

$$
\begin{aligned}
\mu^{(m)}_{\{i\}}(x) &= x+1-v_m(\{i\}) \quad \text{for} \quad x \in [v_m(\{i\})-1, v_m(\{i\})],\\
&= 1-x+v_m(\{i\}) \quad \text{for} \quad x \in [v_m(\{i\}), v_m(\{i\})+1],\\
&= 0 \quad \text{otherwise}.
\end{aligned}
$$

Then it is easy to calculate the values $\beta_m(\mathcal{L}, \boldsymbol{\alpha})$. They are summarized in Table 8.5. To simplify the notations in the head of the table we write

$$
\begin{aligned}
&\beta_m(1) = \beta((I),(1)), \quad \beta_m(2) = \beta((\{1\},\{2\},\{3\}),(1,1,1)),\\
&\beta_m(3) = \beta((\{i\},\{j,k\}),(1,1)),\\
&\beta_m(4) = \beta\left((\{1,2\},\{1,3\},\{2,3\}),\left(\tfrac{1}{2},\tfrac{1}{2},\tfrac{1}{2}\right)\right)
\end{aligned}
$$

for $m = 1,\ldots,15$, $i,j,k \in I$, $i \neq j \neq k \neq i$.

Notice that the game (I, v_4) and (I, w_4) is dealt with in Example 8.3.

□

8.2 Relations to the Fuzzy Core

We already know from Subsection 3.1 that in the deterministic case the balancedness of a coalition game with side-payments is equivalent to the existence of its non-empty core attained by the coalition of all players. This relation is not so strong for the fuzzy extensions of coalition games.

Similarly to the case of convexity, even here we could intuitively expect that even the possibilities of balancedness $\beta(I,w)$ of the game and effectivity $\varepsilon(I)$ of the coalition I would have to fulfill some relation, in the best case $\varepsilon(I) = \beta(I,w)$. In the extremal case it is so.

Let (I,w) be a fuzzy extension of (I,v) such that for any coalition K, $\mu_K(x) = 1$ iff $x = v(K)$. Then we already know (see Chapter 6) that in such a case, we have $\varepsilon(I) = 1$ iff I is deterministically effective in (I,v). This is true iff (I,v) is balanced. In case of this type of a fuzzy extension, this may happen iff $\beta(I,w) = 1$.

A stronger relation, though it would be pleasant, is not fulfilled as we can see in the following example which presents further properties of the 3-players games investigated in the previous Example 8.5.

Example 8.6 Let us consider the 15-element set of coalition games (I, v_m) and their fuzzy extensions (I, w_m) already dealt with in Example 8.5. In

Table 8.1. The possibilities of balancedness in Example 8.5. The values $\beta_m(\mathcal{L}, \alpha)$

m	$v_m(\{i\})$	$\beta_m(1)$	$\beta_m(2)$	$\beta_m(3)$	$\beta_m(4)$	$\beta(I, w_m)$
1	–2	1	1	1	3/4	3/4
2	–1	1	1	1	3/4	3/4
3	–1/2	1	1	1	3/4	3/4
4	0	1	1	1	3/4	3/4
5	1/5	1	1	14/15	3/4	3/4
6	1/2	1	1	5/6	3/4	3/4
7	3/5	1	1	4/5	3/4	3/4
8	3/4	1	1	15/16	3/4	3/4
9	4/5	1	9/10	11/15	3/4	11/15
10	1	1	3/4	2/3	3/4	2/3
11	6/5	1	3/5	3/5	3/4	3/5
12	3/2	1	3/8	1/2	3/4	3/8
13	8/5	1	3/10	7/15	3/4	3/10
14	2	1	0	1/3	3/4	0
15	3	1	0	0	3/4	0

Table 8.2. Possibilities of balancedness and effectivity of games in Example 8.6 (cf. Table 7.4 and Table 8.5)

m	$v_m(\{i\})$	$\beta(I, w_m)$	$\varepsilon_m(I)$	$\gamma(I, w_m)$
1	-2	$3/4 = 0,75$	$3/5$	$3/5$
2	-1	$3/4 = 0,75$	$3/5$	$3/5$
3	$-1/2$	$3/4 = 0,75$	$3/5$	$3/5$
4	0	$3/4 = 0,75$	$3/5$	$3/5$
5	$1/5$	$3/4 = 0,75$	$3/5$	$3/5$
6	$1/2$	$3/4 = 0,75$	$3/5$	$3/5$
7	$3/5$	$3/4 = 0,75$	$3/5$	$3/5$
8	$3/4$	$3/4 = 0,75$	$3/5$	$3/5$
9	$4/5$	$11/15 = 0,733$	$3/5$	$3/5$
10	1	$2/3 = 0,667$	$3/5$	1
11	$6/5$	$3/5 = 0,6$	$3/5$	1
12	$3/2$	$3/8 = 0,375$	$3/8$	1
13	$8/5$	$3/10 = 0,3$	$3/10$	1
14	2	$0 = 0$	0	1
15	3	$0 = 0$	0	1

Table 8.2 we compare the possiblities of their balancedness and of the effectivity of the all-players coalition. For this comparison we have summarized the values presented in Table 7.4 (Example 7.4) and Table 8.5 (Example 8.5).

□

Looking at Table 8.2 we can observe two interesting phenomena. First, $\beta(I, w_m) \geq \varepsilon_m(I)$ for all $m = 1, 2, \ldots, 15$, which appears to offer the validity of an implication due to which the fuzzy effectivity of I implies the fuzzy balancedness. Second, the equality between $\beta(I, w)$ and $\varepsilon(I)$ seems to be fulfilled also for the case where both values vanish to zero. If this rule is fulfilled in general, then there could exist an equivalence between balancedness and non-emptiness of core realized by I at least for the extremal situations of certain fuzzy balancedness (certain non-emptiness of fuzzy core) and certain fuzzy non-balancedness (certain emptiness of the fuzzy core). Unfortunately, there up to now do not exist valid proofs of the general validity of these two properties.

9. Fuzzy Shapley Value

It is evident that the vagueness of coalitional pay-offs implies also the vagueness of such concept like the vector of Shapley values. Its formal re-formulation into the language of fuzzy coalition game theory appears to be obvious and purely mechanical. Anyhow, it is worth some closer attention.

Quite formally, it is possible to re-write formula (3.10) specifying the Shapley value for player $i \in I$ into the terms of fuzzy games and to define fuzzy quantities T_i, $i \in I$, with membership functions $\sigma_i : R \to [0,1]$ by substituting the fuzzy numbers $w(K)$ instead of $v(K)$.

In symbols

$$T_i = \sum\nolimits_{K \subset I}^{\oplus} \frac{(n-k)!\,(k-1)!}{n!}\,(w(K) \oplus (-w(K - \{i\}))) \tag{9.1}$$

where the summation symbol $\sum^{\oplus}$ denotes the addition of fuzzy quantities. Due to (2.3) and (2.7) this procedure is possible and formally correct. Numbers $\sigma_i(x)$ for $x \in R$ determine the possibility with which x is the value of T_i for each $i \in I$.

Formula (9.1) is a formal analogy of (3.10) and it is not guaranteed that it fulfills a fuzzified counterpart of the axioms being demanded for the deterministic Shapley value and remembered in Subsection 3.1.

This problem was solved in [37] and the result is quite optimistic. It was show that:

- The values $\sigma_i(x)$ do not depend on the order of player i in the set I.

- If we denote

  $$T = \sum\nolimits_{i=1}^{\oplus} T_i$$

  (in the sense used in (9.1), if (I,w) is a fuzzy extension of (I,v) and t_i, $i \in I$, are Shapley values of (I,v) then $\sum_{i \in I} t_i$ is the modal value of the fuzzy quantity T.

  It means that $T \succsim w(I)$ and $w(I) \succsim T$ which also means $T \sim w(I)$ in the sense of (2.9) and (2.10).

- If (I, v) and (I, v') are coalitional games and (I, w), (I, w') their fuzzy extensions, if T_i and T'_i, $i \in I$, are respective Shapley values, if we define deterministic game $(I, v + v')$ and its fuzzy extension $(I, w \oplus w')$ such that for any $K \subset I$,

$$(v + v')(K) = v(K) + v'(K), \quad (w \oplus w')(K) = w(K) \oplus w'(K),$$

  and if T^*_i are fuzzy Shapley values of $(I, w \oplus w')$ then

  (a) $t_i + t'_i$ are modal values of T^*_i, $i \in I$,

  (b) if all fuzzy quantities $w(K)$ and $w'(K)$ are almost trapezoidal then also

$$T^*_i \sim_{\oplus} T_i \oplus T'_i,$$

  where t_i, t'_i are Shapley values of (I, v) and (I, v') respectively.

  Let us note that the above property (b) is valid in all cases in which the distributivity law

$$(r_1 + r_2) \cdot a \sim_{\oplus} r_1 \cdot a \oplus r_2 \cdot a, \quad r_1, r_2 \in R, \ a \in I\!R$$

  (see Chapter 2) is fulfilled for all $w(K)$ and $w'(K)$.

The previous properties of fuzzy Shapley values mean that they fulfill, in the sense of the fuzzy set theoretical concepts like addition $\oplus$ and additive equivalence $\sim_{\oplus}$, the axioms formulated by Shapley at least for a significant class of fuzzy coalitional games with almost trapezoidal fuzzy characteristic values $w(K)$.

As the Shapley values in the deterministic game form a vector, it is rational to respect this fact also in the fuzzified case and to define the possibility that a real-valued vector $\boldsymbol{x} = (x_i)_{i \in I}$ is a vector of Shapley values as the number

$$\sigma(\boldsymbol{x}) = \min(\sigma_i(x_i) : i \in I). \tag{9.2}$$

The defining relation (2.3) of the addition operation with fuzzy quantities immediately implies that the vector $(t_i)_{i \in I}$ of deterministic Shapley values in (I, v) is one of possible values of the fuzzy vector $(T_i)_{i \in I}$ and, moreover, for all $i \in I$ there is

$$\sigma_i(t_i) = 1 \text{ which means } \sigma(\boldsymbol{t}) = 1, \text{ where } \boldsymbol{t} = (t_i)_{i \in I}.$$

If the membership functions μ_K, $K \subset I$, are unimodal, i.e. $\mu_K(x) = 1$ iff $x = v(K)$, for all $K \subset I$, then

$$\begin{aligned} \sigma_i(x) &= 1 \quad \text{iff } x = t_i, \ i \in I, \qquad \text{and} \\ \sigma(\boldsymbol{x}) &= 1 \quad \text{iff } \boldsymbol{x} = \boldsymbol{t}. \end{aligned}$$

If (I, w) is superadditive with possibility $\nu_{\text{super}}(i, w) = 1$, then the inequality

$$\sum_{i \in I} t_i \leq v(I)$$

mentioned in Section 3.1 implies – due to (2.9) – that the relation

$$w(I) \succsim \sum_{i \in I}^{\oplus} T_i$$

is valid with the possibility

$$\nu_{\succsim}\left(w(I), \sum_{i=1}^{\oplus} T_i\right) = 1.$$

The fuzzy quantities representing the Shapley value can be illustrated by the following two examples.

Example 9.1 Let us consider the 2-players game (I, v) and its fuzzy extension (I, w) already discussed in Example 6.1 and Example 8.1. It means that $I = \{1, 2\}$, $v(\{1\}) = 1$, $v(\{2\}) = 2$, $v(I) = 5$ and for any $K \subset I$, there is

$$\begin{aligned} \mu_K(x) &= x + 1 - v(K) & &\text{for} & &x \in [v(K) - 1, v(K)], \\ &= 1 + v(K) - x & &\text{for} & &x \in [v(K), v(K) + 1], \\ &= 0 & &\text{otherwise.} \end{aligned}$$

Then

$$\begin{aligned} T_1 = & \frac{1! \cdot 0!}{2!} w(\{1\}) \oplus \frac{0! \cdot 1!}{2!} (w(I) - w(\{1\})) = \\ = & \frac{1}{2} w(\{1\}) \oplus \left(-\frac{1}{2}\right) \cdot w(\{1\}) + \frac{1}{2} w(I), \end{aligned}$$

which means – due to (2.3),(2.4) and (2.7) – that

$$\begin{aligned} \sigma_1(x) &= 2(x-1)/3 & &\text{for} & &x \in [1, 5/2], \\ &= 2(4-x)/3 & &\text{for} & &x \in [5/2, 4], \\ &= 0 & &\text{otherwise.} \end{aligned}$$

Analogously, we have

$$T_2 = \frac{1}{2} w(\{2\}) \oplus \left(-\frac{1}{2}\right) \cdot w(\{2\}) + \frac{1}{2} w(I)$$

$$\begin{aligned} \sigma_2(x) &= 2(x-1)/3 & &\text{for} & &x \in [1, 5/2], \\ &= 2(4-x)/3 & &\text{for} & &x \in [5/2, 1], \\ &= 0 & &\text{otherwise.} \end{aligned}$$

For $\boldsymbol{x} = (x_1, x_2)$, there is

$$\sigma(\boldsymbol{x}) = \min\left(\sigma_1(x_1), \sigma_2(x_2)\right)$$

which implies the following analytic form

$$
\begin{array}{llll}
\sigma(\boldsymbol{x}) & = 1 & \text{for} & \boldsymbol{x} = (5/2, 5/2), \\
\sigma(\boldsymbol{x}) & > 0 & \text{for} & \boldsymbol{x} \text{ from inside of the square} \\
& & & \text{with vertices} \\
& & & (1,1), (1,4), (4,1), (4,4),
\end{array}
$$

$$
\begin{array}{llll}
\sigma(\boldsymbol{x}) & = 2(x_1 - 1)/2 & \text{for} & x_2 \in [1, 5/2],\ x_1 \in [x_2, 5 - x_2], \\
& = 2(4 - x_2)/3 & \text{for} & x_2 \in [5/2, 4],\ x_1 \in [5 - x_2, x_2], \\
& = 2(x_1 - 1)/3 & \text{for} & x_2 \in [1, 5/2],\ x_1 \in [x_1, 5 - x_1], \\
& = 0 & \text{otherwise.} &
\end{array}
$$

It is easy to see that [cf. (2.10)]

$$\nu_{\succsim}(w(I), T_1 \oplus T_2) = \nu_{\sim}(T_1 \oplus T_2, w(I)) = \nu_{\succsim}(w(I), T_1 \oplus T_2) = 1.$$

□

Example 9.2 Rather simpler is the case of a 3-players game. Let us consider the game (I, v) and its fuzzy extension (I, w) already dealt with in Example 8.4, where $v(\{i\}) = 1$, $v(\{i,j\}) = 3$, $v(I) = 5$, for $i, j \in I$, $i \neq j$, and for any $K \subset I$ there is

$$
\begin{array}{llll}
\mu_K(\boldsymbol{x}) & = x + 1 - v(K) & \text{for} & x \in [v(K) - 1, v(K)], \\
& = 1 + v(K) - x & \text{for} & x \in [v(K), v(K) + 1], \\
& = 0 & \text{otherwise.} &
\end{array}
$$

Obviously, $T_1 = T_2 = T_3$, and it is not difficult to verify that for all $i = 1, 2, 3$, $j, k \in I$, $i \neq j \neq k \neq i$, we have

$$
\begin{aligned}
T_i = {} & \frac{1}{3} w(\{i\}) + \frac{1}{6} (w(\{i,j\}) \oplus (-w(\{j\}))) \oplus \frac{1}{6} (w(\{i,k\}) \oplus (-w(\{k\})) \\
& + \frac{1}{3} (w(I) \oplus (-w(\{j,k\})), \quad j \neq i \neq k \neq j,
\end{aligned}
$$

so that

$$
\begin{array}{llll}
\sigma_i(x) & = (3x - 1)/4 & \text{for} & x \in [1/3, 5/3], \\
& = 3(3 - x)/4 & \text{for} & x \in [5/3, 3], \\
& = 0 & \text{otherwise.} &
\end{array}
$$

for every $\boldsymbol{x} = (x_1, x_2, x_3) \in R^3$, and

$$
\begin{aligned}
& \sigma(\boldsymbol{x}) = \min(\sigma_1(x_1), \sigma_2(x_2), \sigma_3(x_2)), \\
& \sigma(\boldsymbol{x}) = 1 \quad \text{iff} \quad \boldsymbol{x} = (5/3, 5/3, 5/3)
\end{aligned}
$$

and $\sigma(\boldsymbol{x}) > 0$ for $\boldsymbol{x}$ from inside the cube limited by the vertices

$$(1/3, 1/3, 1/3),\ (3, 1/3, 1/3),\ (1/3, 3, 1/3),\ (1/3, 1/3, 3),$$
$$(3, 3, 1/3),\ (3, 1/3, 3),\ (1/3, 3, 3),\ (3, 3, 3).$$

If we denote by $\sigma_{1,2,3} : R \to [0, 1]$ the membership function of $T_1 \oplus T_2 \oplus T_3$, then

$$\begin{aligned} \sigma_{1,2,3}(x) &= (x-1)/4 \quad \text{for} \quad x \in [1,5], \\ &= (9-x)/4 \quad \text{for} \quad x \in [5,9], \\ &= 0 \quad \text{otherwise,} \end{aligned}$$

and

$$\nu_{\succsim}(w(I), T_1 \oplus T_2 \oplus T_3) = \nu_{\sim}(w(I), T_1 \oplus T_2 \oplus T_3) = \nu_{\succsim}(T_1 \oplus T_2 \oplus T_3, w(I)) = 1.$$

□

Part III

FUZZY BARGAINING WITHOUT SIDE–PAYMENTS

In this part we focus our attention on the fuzzification of more general type of coalition games without side-payments, briefly recalled in Section 3.2. Formula (3.18) illustrates the fact that the general coalition games without side-payments include the games with side-payments as their special case.

However, this strong bound between the games with and without side-payments turns much weaker if we approach their fuzzification. The main reason for this was already mentioned in Chapter 4. While in the case of games with side-payments we fuzzify numerical quantities, real numbers, determining the total pay-offs of coalitions, in the fuzzification of the general games we operate with fuzzy subsets of multidimensional real space. The processing of these two fuzzy set theoretical objects – fuzzy numbers and fuzzy subsets of multidimensional space – their combination, composition and comparison are of an essentially different nature. It is the main reason why the results presented below are of a type than those formulated in Part II.

Let us consider a coalition game without side-payments $(I, \mathcal{V})$ and let us construct for every coalition $K \subset I$ a fuzzy subset $\mathcal{W}(K)$ of R^I with the membership function $\kappa_K : R^I \to [0,1]$ such that:

- $\kappa_K(\boldsymbol{x}) = 1$, for all $\boldsymbol{x} \in \mathcal{V}(K)$,
- the set $\{\boldsymbol{x} \in R^I : \kappa_K(\boldsymbol{x}) = 1\}$ is closed,
- if $\boldsymbol{x}, \boldsymbol{y} \in R^I$, $x_i \geq y_i$, for all $i \in K$, then $\kappa_K(\boldsymbol{x}) \leq \kappa_K(\boldsymbol{y})$,
- if $K \neq \emptyset$, then there exists $\boldsymbol{x} \in R^I$ such that $\kappa_K(\boldsymbol{x}) = 0$.

The previous conditions, together with (3.13) and (3.14), mean that for every $K \subset I$ there exists an $\boldsymbol{x} \in R^I$ such that $\kappa_K(\boldsymbol{x}) = 1$, and for empty coalition, there is $\kappa_\emptyset(\boldsymbol{x}) = 1$, for all $\boldsymbol{x} \in R^I$. The pair $(I, \mathcal{W})$ is called a *fuzzy extension of the coalition game* $(I, \mathcal{V})$, and the mapping $\mathcal{W}$ connecting each coalition $K \subset I$ with a fuzzy subset of R^I is called a *fuzzy general characteristic function* of $(I, \mathcal{W})$.

The fuzzy extensions of deterministic coalition games without side-payments are a form of *fuzzy coalition games* which are described and investigated in this part of the book.

If $(I, \mathcal{W})$ and $(I, \mathcal{W}')$ are fuzzy extensions of a coalition game without side-payments $(I, \mathcal{V})$, then also $(I, [\mathcal{W} \cup \mathcal{W}'])$ and $(I, [\mathcal{W} \cap \mathcal{W}'])$, where for any $K \subset I$, $[\mathcal{W} \cup \mathcal{W}'], (K) = \mathcal{W}(K) \cup \mathcal{W}'(K)$ and $[\mathcal{W} \cap \mathcal{W}'], (K) = \mathcal{W}(K) \cap \mathcal{W}'(K)$, are also fuzzy extensions of $(I, \mathcal{V})$, as shown in [36].

It can be useful to introduce at least some simple examples of fuzzy extensions of coalition games without side-payments.

Example 9.3 Let us consider a deterministic game $(I, \mathcal{V})$ very similar to that discussed in Example 3.10, with $I = \{1, 2\}$, $\mathcal{V}(\emptyset) = R^2$, and

$$\begin{aligned} \mathcal{V}(\{i\}) &= \{\boldsymbol{x} \in R^2 : x_i \leq 1\}, \quad i = 1, 2, \\ \mathcal{V}(I) &= \{\boldsymbol{x} \in R^2 : x_1 \cdot x_2 \leq 4\}. \end{aligned}$$

It can be fuzzy extended, e.g., into the following fuzzy coalition game $(I, \mathcal{W})$ with $\mathcal{W}(\emptyset) = R^2$, i.e.,

$$
\begin{aligned}
\kappa_{\emptyset}(\boldsymbol{x}) &= 1 && \text{for all} \quad \boldsymbol{x} \in R^2 \\
\kappa_{\{1\}}(\boldsymbol{x}) &= 1 && \text{for} \quad \boldsymbol{x} \in R^2,\ x_1 \le 1, \\
&= 2 - x_1 && \text{for} \quad \boldsymbol{x} \in R^2,\ x_1 \in [1,2], \\
&= 0 && \text{for} \quad \boldsymbol{x} \in R^2,\ x_1 > 2, \\
\kappa_{\{2\}}(\boldsymbol{x}) &= 1 && \text{for} \quad \boldsymbol{x} \in R^2,\ x_2 \le 1, \\
&= 3/2 - x_2/2 && \text{for} \quad \boldsymbol{x} \in R^2,\ x_1 \in [1,3], \\
&= 0 && \text{for} \quad \boldsymbol{x} \in R^2,\ x_3 > 3, \\
\kappa_I(\boldsymbol{x}) &= 1 && \text{for} \quad \boldsymbol{x} \in R^2,\ x_1 \cdot x_2 \le 4, \\
&= 2/3 && \text{for} \quad \boldsymbol{x} \in R^2,\ x_1 \cdot x_2 \in (4,9], \\
&= 1/3 && \text{for} \quad \boldsymbol{x} \in R^2,\ x_1 \cdot x_2 \in (9,16], \\
&= 0 && \text{for} \quad \boldsymbol{x} \in R^2,\ x_1 \cdot x_2 > 16.
\end{aligned}
$$

□

Example 9.4 The coalition game $(I, \mathcal{V})$ dealt with in Example 3.9 with

$$
\begin{aligned}
\mathcal{V}(I) &= \{\boldsymbol{x} \in R^2 : x_1 + x_2 \le 1\} \cap \{\boldsymbol{x} \in R_1^2 : x_2 \le 1\} \\
\mathcal{V}(\{1\}) &= \{\boldsymbol{x} \in R^2 : x_1 \le \sqrt{3}/3\}, \\
\mathcal{V}(\{2\}) &= \{\boldsymbol{x} \in R^2 : x_2 \le \sqrt{6}/3\}.
\end{aligned}
$$

can be fuzzy extended into the following game $(I, \mathcal{W})$ with

$$
\begin{aligned}
\kappa_{\emptyset}(\boldsymbol{x}) &= 1 \quad \text{for all } \boldsymbol{x} \in R^2, \\
\kappa_{\{1\}}(\boldsymbol{x}) &= 1 \quad \text{for } \boldsymbol{x} \in R^2,\ x_1 \le \sqrt{3}/3, \\
&= 0 \quad \text{for } \boldsymbol{x} \in R^2,\ x_1 > \sqrt{3}/3,
\end{aligned}
$$

(the set $\mathcal{V}(\{1\})$ has not been, in fact, fuzzified and it keeps its deterministic nature), and

$$
\begin{aligned}
\kappa_{\{2\}}(\boldsymbol{x}) &= 1 && \text{for} \quad \boldsymbol{x} \in R^2,\ x_2 \le \sqrt{6}/3, \\
&= 1 + \sqrt{6}/3 - x && \text{for} \quad \boldsymbol{x} \in R^2,\ x_2 \in (\sqrt{6}/3,\ 1 + \sqrt{6}/3), \\
&= 0 && \text{for} \quad \boldsymbol{x} \in R^2,\ x_2 \ge 1 + \sqrt{6}/3,
\end{aligned}
$$

and the membership function κ_I can be defined as follows

$$\begin{array}{rclll} \kappa_I(\boldsymbol{x}) & = & 1 & \text{for} & \boldsymbol{x} \in \mathcal{V}(I), \\ \kappa_I(\boldsymbol{x}) & = & 1-\lambda & \text{for} & \boldsymbol{x} \in R^2, x_1^2 + x_2^2 = (1+\lambda)^2, \\ & & & & x_1 \geq 0,\ x_2 \geq 0, \\ & & & \text{or} & \boldsymbol{x} \in R^2,\ x_1 = 1+\lambda,\ x_2 < 0, \\ & & & \text{or} & \boldsymbol{x} \in R^2,\ x_2 = 1+\lambda,\ x_1 < 0, \\ & & & \text{and} & \lambda \in [0,1], \end{array}$$

and

$$\begin{array}{rclll} \kappa_I(\boldsymbol{x}) & = & 0 & \text{for} & \boldsymbol{x} \in R^2,\ x_1^2 + x_2^2 > 4,\ x_1 \geq 0,\ x_2 \geq 0, \\ & & & \text{or} & \boldsymbol{x} \in R^2,\ x_1 > 2,\ x_2 < 0, \\ & & & \text{or} & \boldsymbol{x} \in R^2,\ x_2 > 2,\ x_1 < 0. \end{array}$$

This extension consequently distinguishes between imputations from $\mathcal{V}(I)$ which are taken for surely accessible and the others whose accessibility regularly decreases with the distance from $\mathcal{V}(I)$ and vanishes if the distance is 1 or more.

□

Example 9.5 The deterministic coalition game without side-payment $(I, \mathcal{V})$ considered in the previous Example 9.4 can be fuzzy extended in a less rigorous ways.

Its fuzzy extension $(I, \mathcal{W})$ is characterized by the following characteristic functions: $\kappa_\emptyset$, $\kappa_{\{1\}}$ and $\kappa_{\{2\}}$ are identical with the respective membership functions given in Example 9.4. The function κ_I is defined in the following way

$$\begin{array}{rcll} \kappa_I(\boldsymbol{x}) & = & 1 & \text{for } \boldsymbol{x} \in R^2,\ x_1 \leq 1 \text{ and } x_2 \leq 1, \\ \kappa_I(\boldsymbol{x}) & = & 0 & \text{for } \boldsymbol{x} \in R^2,\ x_1 > 1 \text{ or } x_2 > 2, \\ \kappa_I(\boldsymbol{x}) & = & 1-\lambda & \text{for } \boldsymbol{x} \in R^2,\ x_1 = 1+\lambda,\ x_2 \leq 1+\lambda, \\ & & & \text{or } \boldsymbol{x} \in R^2,\ x_2 = 1+\lambda,\ x_1 \leq 1+\lambda, \\ & & & \text{and } \lambda \in [0,1]. \end{array}$$

This $(I, \mathcal{W})$ is a fuzzy extension of $(I, \mathcal{V})$, too.

□

10. Fuzzy Superoptimum

In the deterministic coalition games we have introduced the concept of a superoptimum (see (3.16)) to determine the sets of imputations which cannot be dominated by any imputation accessible by the coalition. The domination relation is the basic tool not only for the introduction of a superoptimum but, consequently, also of the subadditivity, core and some other concepts. This means that a transition to general fuzzy coalition games is a very important step in the process of fuzzification of general coalition games.

It appears rational to respect the possibility of imputations entering the domination relation and to demand the possibility of the dominating imputation to be at least as big as the possibility of the dominated one, where the possibility means the value of the membership function κ_K of the set $\mathcal{W}(K)$ of the coalition to which the domination is related. If we respect this principle then the possibility of the validity of domination via a coalition is equal to the possibility of the dominating imputation or it vanishes with respect to the relation between possibilities of both imputations entering the relation.

The previous heuristic principle can be rather arguable but there are good reasons for its application, as we will see in Chapter 14. Its formal presentation is as follows.

Let us consider a fuzzy extension (I, W) of a coalition game $(I, \mathcal{V})$, let $K \subset I$ be a coalition and let $\boldsymbol{x}, \boldsymbol{y} \in R^I$ be imputations. Then we say that $\boldsymbol{x}$ *significantly dominates* $\boldsymbol{y}$ *via coalition* K iff $\boldsymbol{x} \operatorname{dom}_K \boldsymbol{y}$ in the sense of (3.15) and $\kappa_K(\boldsymbol{x}) \geq \kappa_K(\boldsymbol{y})$.

If $\boldsymbol{x} \operatorname{dom}_K \boldsymbol{y}$, then the possibility that this domination is significant is, by definition, equal to the number $\kappa_K(\boldsymbol{x})$ iff $\kappa_K(\boldsymbol{x}) \geq \kappa_K(\boldsymbol{y})$, and it is equal to 0 iff the opposite inequality $\kappa_K(\boldsymbol{x}) < \kappa_K(\boldsymbol{y})$ is fulfilled. In a very formal terminology, the relation of significant domination (or the property of two imputations and a coalition to dominate significantly via the coalition) can be identified with a fuzzy subset of the class of ordered triples

$$\{(\boldsymbol{x}, \boldsymbol{y}, K) : \boldsymbol{x} \operatorname{dom}_K \boldsymbol{y}\}.$$

The membership function of this fuzzy subset will be denoted $\nu_K(\boldsymbol{x}, \boldsymbol{y})$, where

$$\begin{aligned}\nu_K(\boldsymbol{x},\boldsymbol{y}) &= \kappa_K(\boldsymbol{x}) \quad \text{for } \kappa_K(\boldsymbol{x}) \geq \kappa_K(\boldsymbol{y}),\ \boldsymbol{x}\,\mathrm{dom}_K\boldsymbol{y} \\ &= 0 \quad \text{for } \kappa_K(\boldsymbol{x}) < \kappa_K(\boldsymbol{y}),\ \boldsymbol{x}\,\mathrm{dom}_K\boldsymbol{y}, \\ &= 0 \quad \text{for } \boldsymbol{x}, \text{ with } \mathrm{dom}_K\boldsymbol{y} \text{ not fulfilled,}\end{aligned} \tag{10.1}$$

for $\boldsymbol{x}, \boldsymbol{y} \in R^I$, $K \subset I$.

The properties of the fuzzy general characteristic function obviously imply that if $\boldsymbol{x}$ *significantly dominates* $\boldsymbol{y}$ *via* K with a positive possibility $\nu_K(\boldsymbol{x},\boldsymbol{y})$, then $\kappa_K(\boldsymbol{x}) = \kappa_K(\boldsymbol{y})$ as the inequality $\kappa_K(\boldsymbol{x}) \geq \kappa_K(\boldsymbol{y})$ follows from the significancy of the domination, and $\kappa_K(\boldsymbol{y}) \geq \kappa_K(\boldsymbol{x})$ follows from the definitional properties of $\mathcal{W}(K)$ and κ_K.

Having introduced the concept of a significant domination, we can define a *fuzzy superoptimum* of the coalition K as a fuzzy subset $\mathcal{W}^*(K)$ of imputations which are not significantly dominated by any imputation from $\mathcal{W}(K)$.

Since $\mathcal{W}(K)$ is a fuzzy set and a significant domination via K is a fuzzy relation (it means that also its opposite is connected with some complementary possibility), then $\mathcal{W}^*(K)$ is naturally a fuzzy subset of R^I with the membership function $\kappa_K^* : R^I \to [0,1]$, where

$$\kappa_K^*(\boldsymbol{y}) = 1 - \sup_{\boldsymbol{x}\in R^I} \nu_K(\boldsymbol{x},\boldsymbol{y}) = \inf_{\boldsymbol{x}\in R^I} (1 - \nu_K(\boldsymbol{x},\boldsymbol{y})). \tag{10.2}$$

Properties of the membership functions κ_K combined with (10.2) imply that for an empty coalition, $\kappa_\emptyset^*(\boldsymbol{x}) = 1$, for all $\boldsymbol{x} \in R^I$, and for any non-empty coalition $K \subset I$, $K \neq \emptyset$, there exists $\boldsymbol{x}_0 \in R^I$ such that $\kappa_K^*(\boldsymbol{x}_0) = 1$ and, moreover, $\kappa_K^*(\boldsymbol{y}) = 1$ for all $\boldsymbol{y} \in R^I$ such that $\boldsymbol{y}\,\mathrm{dom}_K\,\boldsymbol{x}_0$.

It is also easy to see that if $\chi_K^* : R^I \to \{0,1\}$ is the characteristic function of the set $\mathcal{V}^*(K)$ (it means $\chi_K^*(\boldsymbol{x}) = 1$ for $\boldsymbol{x} \in \mathcal{V}^*(K)$ and $\chi_K^*(\boldsymbol{x}) = 0$ for $\boldsymbol{x} \notin V^*(K)$), then $\chi_K^*(\boldsymbol{x}) \geq \kappa_K^*(\boldsymbol{x})$, for all $\boldsymbol{x} \in R^I$. Other properties of superoptima are not very systematic in general. Even if in the deterministic game for any $\boldsymbol{x} \in \mathcal{V}^*(K)$ and $\boldsymbol{y} \in R^I$ such that $\boldsymbol{y}\,\mathrm{dom}_K\,\boldsymbol{x}$, $\boldsymbol{y} \in \mathcal{V}^*(K)$, the fuzzy extensions $(I,\mathcal{W})$ of $(I,\mathcal{V})$ may, but need not posses, a fuzzy generalization of this property. Namely, if $\boldsymbol{x}, \boldsymbol{y} \in R^I$ and $\boldsymbol{y}\,\mathrm{dom}_K\,\boldsymbol{x}$, then $\kappa_K^*(\boldsymbol{y})$ can be in some cases also smaller than $\kappa_K^*(\boldsymbol{x})$ as shown in the following examples. In the first one of them the monotonicity of κ_K^* is preserved, in the second one it is not.

The following examples also show how to define a *fuzzy Pareto optimum* of the coalition $K \subset I$. Analogously to the deterministic case, it is the intersection of fuzzy sets $\mathcal{W}(K)$ and $\mathcal{W}^*(K)$ which we denote

$$\mathcal{Q}(K) = \mathcal{W}(K) \cap \mathcal{W}^*(K). \tag{10.3}$$

It means that the membership function $\rho_K : R^I \to [0,1]$ of $\mathcal{Q}(K)$ is defined by

$$\rho_K(\boldsymbol{x}) = \min(\kappa_K(\boldsymbol{x}), \kappa_K^*(\boldsymbol{x})), \quad \boldsymbol{x} \in R^I. \tag{10.4}$$

Properties of the Pareto optimum are deeply connected with the properties of a superoptimum $\mathcal{W}^*(K)$ and, in general, the following ones seem to be the most significant.

If for a coalition $K \subset I$, $K \neq \emptyset$, $\kappa_K(\boldsymbol{x}) = 1$ iff $\boldsymbol{x} \in \mathcal{V}(K)$, then it can be easily seen that $\kappa_K^*(\boldsymbol{x}) = \rho(\boldsymbol{x}) = 1$, for all $\boldsymbol{x} \in \mathcal{P}(K) = \mathcal{V}(K) \cap \mathcal{V}^*(K)$, as $\kappa(\boldsymbol{y}) < 1$ for all $\boldsymbol{y} \in R^I$ such that $\boldsymbol{y} \operatorname{dom}_K \boldsymbol{x}$, $\kappa_K^*(\boldsymbol{x}) = 1$.

The previous statement can be generalized. If $(I, \mathcal{W})$ is a fuzzy extension of $(I, \mathcal{V})$ and if we denote, for $K \subset I$,

$$\overline{\mathcal{V}}(K) = \{\boldsymbol{x} \in R^I : \kappa_K(\boldsymbol{x}) = 1\},$$

then $\overline{\mathcal{V}}(K)$ is closed, the pair $(I, \overline{\mathcal{V}})$ forms a deterministic coalition game, and $(I, \mathcal{W})$ is also a fuzzy extension of $(I, \overline{\mathcal{V}})$.

It is possible to construct $\overline{\mathcal{V}}^*(K)$ for any $K \subset I$ using (3.16), and $\overline{\mathcal{P}}(K) = \overline{\mathcal{V}}(K) \cap \overline{\mathcal{V}}^*(K)$. If $K \neq \emptyset$, then $\kappa_K^*(\boldsymbol{x}) = 1$ for all $\boldsymbol{x} \in \overline{\mathcal{P}}(K)$. Moreover,

$$\overline{\mathcal{V}}^*(K) = \{\boldsymbol{x} \in R^I : \kappa_K^*(\boldsymbol{x}) > 0\}.$$

If $(I, \mathcal{W})$ is a fuzzy extension of $(I, \mathcal{V})$ and if for $K \subset I$, $\kappa_K(\boldsymbol{x})$ is strictly decreasing for any $\boldsymbol{x}$ from the set

$$\{\boldsymbol{x} \in R : \kappa_K(\boldsymbol{x}) \in (0, 1)\},$$

i.e. $\kappa_K(\boldsymbol{x}) > \kappa_K(\boldsymbol{y})$ if $\boldsymbol{y} \operatorname{dom}_K \boldsymbol{x}$ in the sense of (3.15), then $\kappa_K^*(\boldsymbol{x}) \in \{0, 1\}$.

In particular, if

$$\{\boldsymbol{x} \in R^I : \kappa_K(\boldsymbol{x}) = 1\} = \mathcal{V}(K) \tag{10.5}$$

then

$$\kappa_K^*(\boldsymbol{x}) = 0 \quad \text{for } \boldsymbol{x} \in \mathcal{V}(K) - \mathcal{P}(K), \qquad \kappa_K^*(\boldsymbol{x}) = 1 \quad \text{for } \boldsymbol{x} \in \mathcal{V}^*(K)$$

and also $\rho_K = 1$ for $\boldsymbol{x} \in \mathcal{P}(K)$ and $\rho_K(\boldsymbol{x}) = \kappa_K(\boldsymbol{x})$ for $\boldsymbol{x} \notin \mathcal{V}(K)$ (where $\rho_K(\boldsymbol{x})$ is the membership function (10.4)).

Generally, for any $K \neq \emptyset$,

$$\{\boldsymbol{x} \in R^I : \kappa_K(\boldsymbol{x}) < 1\} \subset \{\boldsymbol{x} \in R^I : \kappa_K^*(\boldsymbol{x}) > 0\} \tag{10.6}$$

and

$$\{\boldsymbol{x} \in R^I : \rho_K(\boldsymbol{x}) = 1\} = \{\boldsymbol{x} \in R^I : \kappa_K^*(\boldsymbol{x}) > 0\} - \{\boldsymbol{x} \in R^I : \kappa_K(\boldsymbol{x}) < 1\}. \tag{10.7}$$

The previous results can be interpreted in the following sense. If $(I, \mathcal{W})$ is a natural fuzzy extension of $(I, \mathcal{V})$ in the sense of (10.5) for any coalition $K \subset I$, then also the fuzzy Pareto optimum $\mathcal{Q}(K)$ is a fuzzy extension of $\mathcal{P}(K)$. If the vagueness connected with the accessibility of imputations strictly increases with the distance from the "guaranteed" set $\mathcal{V}(K)$ (i.e.

their possibility decreases), then superoptima of coalitions are identical with superoptima in the original crisp version of the considered game. In such a case also the fuzzy Pareto optima copy the vagueness of fuzzy sets of the accessible imputations $\mathcal{W}(K)$.

The clearest illustration of the above concepts of significant dominating, fuzzy superoptima and fuzzy Pareto optima is the one using the simplest cases of coalitions containing only one or two players. Let us show some of them.

Example 10.1 Let us consider a coalition game $(I, \mathcal{V})$ and its 1-player coalition $\{i\}$ with

$$\mathcal{V}(\{i\}) = \{\boldsymbol{x} \in R^I : x_i \leq 2\}.$$

One of its possible fuzzy extensions is represented by a fuzzy set $\mathcal{W}(\{i\})$ such that

$$\begin{aligned}\kappa_{\{i\}}(\boldsymbol{x}) &= 1 && \text{for} \quad \boldsymbol{x} \in \{\boldsymbol{y} \in R^I : y_i \leq 2\} = \mathcal{V}(\{i\}),\\ &= 3 - x_i && \text{for} \quad \boldsymbol{x} \in \{\boldsymbol{y} \in R^I : 2 < y_i \leq 3\},\\ &= 0 && \text{for} \quad \boldsymbol{x} \in \{\boldsymbol{y} \in R^I : y_i > 3\}.\end{aligned}$$

It is easy to verify that for every $\boldsymbol{x} \in R^I$, $x_i < 2$ there exists $\boldsymbol{y} \in R^I$ such that $y_i = 2$, $\kappa_{\{i\}}(\boldsymbol{y}) = 1 = \kappa_{\{i\}}(\boldsymbol{x})$ and $\boldsymbol{y} \operatorname{dom}_{\{i\}} \boldsymbol{x}$.

For any $\boldsymbol{z} \in R^I$ such that $z_i > 2$ the relation $\boldsymbol{z} \operatorname{dom}_{\{i\}} \boldsymbol{x}$ holds but it is not significant. It means $\nu_{\{i\}}(\boldsymbol{y}, \boldsymbol{x}) = 1$, $\nu_{\{i\}}(\boldsymbol{z}, \boldsymbol{x}) = 0$ and, due to (10.2), $\kappa^*_{\{i\}}(\boldsymbol{x}) = 0$.

Similarly, for $\boldsymbol{y} \in R^I$ such that $y_i \in [2, 3)$ and every $\boldsymbol{z} \in R^I$ such that $z_i > y_i$, also $\kappa_{\{i\}}(\boldsymbol{z}) < \kappa_{\{i\}}(\boldsymbol{y})$. It means that $\boldsymbol{z} \operatorname{dom}_{\{i\}} \boldsymbol{y}$ but not significantly, hence, $\nu_{\{i\}}(\boldsymbol{z}, \boldsymbol{y}) = 0$ and $\kappa^*_{\{i\}}(\boldsymbol{y}) = 1$.

Finally, for $\boldsymbol{y}$ such that $y_i \geq 3$ every $\boldsymbol{z} \in R^I$ such that $z_i > y_i$ dominates $\boldsymbol{y}$ significantly $(\kappa_{\{i\}}(\boldsymbol{z}) = \kappa_{\{i\}}(\boldsymbol{y}))$ but $\nu_{\{i\}}(\boldsymbol{z}, \boldsymbol{y}) = 0$ as $\kappa_{\{i\}}(\boldsymbol{z}) = 0$. Then $\kappa^*_{\{i\}}(\boldsymbol{y}) = 1$.

Summarizing these relations, we obtain

$$\begin{aligned}\kappa^*_{\{i\}}(\boldsymbol{x}) &= 0 \quad \text{for} \quad \boldsymbol{x} \notin \mathcal{V}^*(\{i\})\\ &= 1 \quad \text{for} \quad \boldsymbol{x} \in \mathcal{V}^*(\{i\})\end{aligned}$$

where $\mathcal{V}^*(\{i\}) = \{\boldsymbol{x} \in R^I : x_i \geq 2\}$.

Consequently, the fuzzy Pareto optimum $\mathcal{Q}(\{i\}) = \mathcal{W}(\{i\}) \cap \mathcal{W}^*(\{i\})$ is identified with the membership function

$$\begin{aligned}\rho_{\{i\}}(\boldsymbol{x}) &= 0 && \text{for} \quad \boldsymbol{x} \in \{\boldsymbol{y} \in R^I : y_i < 2 \text{ or } y_i \geq 3\},\\ &= 3 - x_i && \text{for} \quad \boldsymbol{x} \in \{\boldsymbol{y} \in R^I : 2 \leq y_i < 3\}.\end{aligned}$$

In this case, as

$$\{\boldsymbol{x} \in R^I : \kappa_{\{i\}}(\boldsymbol{x}) = 1\} = \mathcal{V}(\{i\}),$$

then $\mathcal{Q}(\{i\})$ is the fuzzy extension of the set $P(\{i\}) = \mathcal{V}(\{i\}) \cap \mathcal{V}^*(\{i\})$ (cf. (3.17)).

□

Example 10.2 Let us consider the coalition game $(I, \mathcal{V})$ already dealt with in the previous Example 10.1, and its 1-player coalition $\{i\}$, $i \in I$, with

$$\mathcal{V}(\{i\}) = \{\boldsymbol{x} \in R^I : x_1 \leq 2\},$$

also processed above.

Let us suppose, in this case, that a fuzzy extension $\mathcal{W}(\{i\})$ of $\mathcal{V}(\{i\})$ is such that

$$\begin{aligned} \kappa_{\{i\}}(\boldsymbol{x}) &= 1 && \text{for} \quad x_i \leq 3 \\ &= \tfrac{4}{x_i - 1} - 1 && \text{for} \quad 3 < x_i < 5 \\ &= 0 && \text{for} \quad x_i \geq 5. \end{aligned}$$

Evidently $\mathcal{V}(\{i\}) \neq \{\boldsymbol{x} \in R^I : \kappa_{\{i\}}(\boldsymbol{x}) = 1\}$, and it is not difficult to verify that (due to the strict monotonicity of $\kappa_{\{i\}}(\boldsymbol{x})$ in x_i at the interval $(3, 5)$) that

$$\begin{aligned} \kappa^*_{\{i\}}(\boldsymbol{x}) &= 1 \quad \text{for} \quad \boldsymbol{x} \in \{\boldsymbol{y} \in R^I : y_i \geq 3\}, \\ &= 0 \quad \text{for} \quad \boldsymbol{x} \in \{\boldsymbol{y} \in R^I : y_i < 3\}. \end{aligned}$$

It means that

$$\begin{aligned} \rho_{\{i\}}(\boldsymbol{x}) &= 0 && \text{for} \quad \boldsymbol{x} \in \{\boldsymbol{y} \in R^I : y_i < 3\} \cup \{\boldsymbol{y} \in R^I : y_i > 5\}, \\ &= \tfrac{4}{x_i - 1} - 1 && \text{for} \quad \boldsymbol{x} \in \{\boldsymbol{y} \in R^I : 3 \leq y_i \leq 5\}. \end{aligned}$$

And the fuzzy Pareto optimum $\mathcal{Q}(\{i\})$ is completely disjoint from its deterministic counterpart

$$P(\{i\}) = \{\boldsymbol{x} \in R^I : x_i = 2\}.$$

□

The shape of $\mathcal{W}^*(\{i\})$ and $\kappa^*_{\{i\}}$ becomes less simple if $\kappa_{\{i\}}(\boldsymbol{x})$ is not strictly monotone for $\boldsymbol{x} \in R^I$ such that $\kappa_{\{i\}}(\boldsymbol{x}) \notin \{0, 1\}$.

Example 10.3 Let us start with a simple step function $\kappa_{\{i\}}$. We consider the coalition game $(I, \mathcal{V})$ already discussed in the previous two examples and the 1-player coalition $\{i\}$ with $\mathcal{V}(\{i\}) = \{\boldsymbol{x} \in R^I : x_i \leq 2\}$.

Let us consider a fuzzy extension $\mathcal{W}(\{i\})$ of $\mathcal{V}(\{i\})$ with

$$\begin{array}{rcll} \kappa_{\{i\}}(\boldsymbol{x}) & = & 1 & \text{for} \quad \boldsymbol{x} \in \{\boldsymbol{y} \in R^I : y_i \le 2\} \\ & = & 1/2 & \text{for} \quad \boldsymbol{x} \in \{\boldsymbol{y} \in R^I : 2 < y_i \le 3\} \\ & = & 0 & \text{for} \quad \boldsymbol{x} \in \{\boldsymbol{y} \in R^I : y_i > 3\}. \end{array}$$

Then, it is easy to verify that

$$\begin{array}{ll} \nu_{\{i\}}(\boldsymbol{x},\boldsymbol{y}) = 1 & \text{for} \quad x_i < 2,\ y_i = 2, \\ \nu_{\{i\}}(\boldsymbol{x},\boldsymbol{y}) = 0 & \text{for} \quad x_i = 2,\ y_i > 2, \\ \nu_{\{i\}}(\boldsymbol{x},\boldsymbol{y}) = 1/2 & \text{for} \quad x_i \in (2,3),\ y_i = 3, \\ \nu_{\{i\}}(\boldsymbol{x},\boldsymbol{y}) = 0 & \text{for} \quad x_i = 3,\ y_i > 3, \\ \nu_{\{i\}}(\boldsymbol{x},\boldsymbol{y}) = 0 & \text{for} \quad x_i > 3,\ y_i > x_i, \end{array}$$

and, consequently,

$$\begin{array}{rclrcll} \kappa^*_{\{i\}}(\boldsymbol{x}) & = & 0, & \rho_{\{i\}}(\boldsymbol{x}) & = & 0 & \text{for } x_i < 2, \\ & = & 1, & & = & 1 & \text{for } x_i = 2, \\ & = & 1/2, & & = & 1/2 & \text{for } 2 < x_i < 3, \\ & = & 1, & & = & 1/2 & \text{for } x_i = 3, \\ & = & 1, & & = & 0, & \text{for } x_i > 3. \end{array}$$

□

The fuzzy extension $\mathcal{W}(\{i\})$ of $\mathcal{V}(\{i\})$ in the following example is more complicated.

Example 10.4 Let us consider a coalition game $(I, \mathcal{V})$ and its 1-player coalition $\{i\}$ already mentioned in the previous examples, i.e. $\mathcal{V}(\{i\}) = \{\boldsymbol{x} : x_i \le 2\}$.

Let $\mathcal{W}(\{i\})$ be fuzzy extension of $\mathcal{V}(\{i\})$ such that

$$\begin{array}{rcll} \kappa_{\{i\}}(\boldsymbol{x}) & = & 1 & \text{for} \quad x_i \le 2, \\ & = & \frac{5}{3} - \frac{1}{3}x_i & \text{for} \quad 2 < x_i \le 3, \\ & = & 1/3 & \text{for} \quad 3 < x_i \le 4, \\ & = & 0 & \text{for} \quad x_i > 4. \end{array}$$

Then, it is possible to compute, using the above formulas, that

$$\begin{array}{rclrcll} \kappa^*_{\{i\}}(\boldsymbol{x}) & = & 0, & \rho_{\{i\}}(\boldsymbol{x}) & = & 0, & \text{for } x_i < 2, \\ & = & 1, & & = & 1 & \text{for } x_i = 2, \\ & = & 1, & & = & \frac{5}{3} - \frac{1}{3}x_i & \text{for } 2 < x_i \le 3, \\ & = & 2/3, & & = & 1/3 & \text{for } 3 < x_i < 3, \\ & = & 1, & & = & 1/3 & \text{for } x_i = 4, \\ & = & 1, & & = & 0 & \text{for } x_i > 4. \end{array}$$

□

Even small changes in the boundary points of the intervals of continuity can influence the values of a superoptimum and Pareto optimum. The comparison of results of the following two examples illustrates this fact. In both of them we deal with fuzzy extensions of the set $\mathcal{V}(\{i\})$, where $i \in I$.

Example 10.5 Let us consider a coalition game $(I, \mathcal{V})$ and a fuzzy extension $\mathcal{W}(\{i\})$ of $\mathcal{V}(\{i\})$ such that, for $\boldsymbol{x} \in R^I$, $\boldsymbol{x} = (x_j)_{j \in I}$, there is

$$\begin{aligned} \kappa_{\{i\}}(\boldsymbol{x}) &= 1 && \text{for} \quad x_i \leq 2, \\ &= 3/4 && \text{for} \quad x_i \in (2,3], \\ &= 1/2 && \text{for} \quad x_i \in (3,4], \\ &= 1/4 && \text{for} \quad x_i \in (4,5], \\ &= 0 && \text{for} \quad x_i > 3. \end{aligned}$$

Then, it is not difficult to verify that

$$\begin{aligned} \kappa^*_{\{i\}}(\boldsymbol{x}) &= 0, & \rho_{\{i\}}(\boldsymbol{x}) &= 0, & &\text{for } x_i < 2, \\ &= 1, & &= 1 & &\text{for } x_i = 2, \\ &= 1/4, & &= 1/4 & &\text{for } x_i \in (2,3), \\ &= 1, & &= 3/4 & &\text{for } x_i = 3, \\ &= 1/2, & &= 1/2 & &\text{for } x_i \in (3,4), \\ &= 1, & &= 1/2 & &\text{for } x_i = 4, \\ &= 3/4, & &= 1/4 & &\text{for } x_i \in (4,5), \\ &= 1, & &= 1/4 & &\text{for } x_i = 5, \\ &= 1, & &= 0 & &\text{for } x_i > 5. \end{aligned}$$

□

A small modification of the values of $\kappa_{\{i\}}$ leads to some changes of $\kappa^*_{\{i\}}$ and $\rho_{\{i\}}$.

Example 10.6 Let $\mathcal{W}(\{i\})$ be a fuzzy extension of $\mathcal{V}(\{i\})$, where $i \in I$ and $(I, \mathcal{V})$ is a coalition game such that for any $\boldsymbol{x} \in R^I$

$$\begin{aligned} \kappa_{\{i\}}(\boldsymbol{x}) &= 1 && \text{for} \quad x_i \leq 2, \\ &= 3/4 && \text{for} \quad x_i \in (2,3), \\ &= 1/2 && \text{for} \quad x_i \in [3,4), \\ &= 1/4 && \text{for} \quad x_i \in [4,5), \\ &= 0 && \text{for} \quad x_i \geq 5. \end{aligned}$$

Then, it is easy to compute, for any $\boldsymbol{x} \in R^I$,

$$
\begin{aligned}
\kappa^*_{\{i\}}(\boldsymbol{x}) &= 0, & \rho_{\{i\}}(\boldsymbol{x}) &= 0, & &\text{for } x_i < 2,\\
&= 1, & &= 1 & &\text{for } x_i = 2,\\
&= 1/4, & &= 1/4 & &\text{for } x_i \in (2,3),\\
&= 1/2, & &= 1/2 & &\text{for } x_i \in [3,4),\\
&= 3/4, & &= 1/4 & &\text{for } x_i \in [4,5),\\
&= 1, & &= 0 & &\text{for } x_i \geq 5.
\end{aligned}
$$

□

For illustration, we present two examples in which a 2-players coalition is treated. Notice a certain similarity between them and Example 10.1 and Example 10.3.

Example 10.7 Let us consider a coalition game $(I, \mathcal{V})$ and one of its 2-players coalitions $K = \{i, j\} \subset I$. Let the set $\mathcal{V}(K)$ be defined as

$$
\begin{aligned}
\mathcal{V}(K) &= \{\boldsymbol{x} \in R^I : x_i + x_j \leq 5\} \cap\\
&\cap \{\boldsymbol{x} \in R^I : x_i \leq 5\} \cap \{\boldsymbol{x} \in R^I : x_j \leq 5\}.
\end{aligned}
$$

To simplify the following notations we denote by $\mathcal{U}(K) \subset R^I$ the set of imputations

$$
\begin{aligned}
\mathcal{U}(K) &= \{\boldsymbol{x} \in R^I : x_i > 5\} \cup \{\boldsymbol{x} \in R^I : x_j > 5\} \cup\\
&\cup \{\boldsymbol{x} \in R^I : x_i = 5 \text{ and } x_j < 0\} \cup\\
&\cup \{\boldsymbol{x} \in R^I : x_j = 5 \text{ and } x_i < 0\}.
\end{aligned}
$$

Let $\mathcal{W}(K)$ be a fuzzy extension of $\mathcal{V}(K)$ such that

$$
\begin{aligned}
\kappa_K(\boldsymbol{x}) &= 1 & &\text{for } \boldsymbol{x} \notin \mathcal{U}(K)\\
& & &\quad \text{and } \boldsymbol{x} \in \{\boldsymbol{y} \in R^I : y_i + y_j \leq 5\}\\
&= 6 - (x_i + x_j) & &\text{for } \boldsymbol{x} \notin \mathcal{U}(K)\\
& & &\quad \text{and } \boldsymbol{x} \in \{\boldsymbol{y} \in R^I : 5 < x_i + x_j \leq 6\}\\
&= 0 & &\text{otherwise.}
\end{aligned}
$$

Then, we can verify that

$$
\begin{aligned}
\kappa^*_K(\boldsymbol{x}) &= 1 & &\text{for } \boldsymbol{x} \in \mathcal{U}(K)\\
& & &\quad \text{or } \boldsymbol{x} \in \{\boldsymbol{y} \in R^I : y_i + y_j \geq 5\}\\
&= 0 & &\text{for } \boldsymbol{x} \notin \mathcal{U}(K)\\
& & &\quad \text{and } \boldsymbol{x} \in \{\boldsymbol{y} \in R^I : y_i + y_j < 5\},
\end{aligned}
$$

which means that $\kappa^*_K(\boldsymbol{x}) = 1$ iff $\boldsymbol{x} \in \mathcal{V}^*(K)$ and $\kappa^*_K(\boldsymbol{x}) = 0$ iff $\boldsymbol{x} \notin \mathcal{V}^*(K)$.
Then, evidently,

$$
\begin{aligned}
\rho_K(\boldsymbol{x}) &= 0 && \text{for} \quad \boldsymbol{x} \in \mathcal{U}(K) \cup \{\boldsymbol{y} \in R^I : y_i + y_j < 5\} \cup \\
& && \qquad \cup \{\boldsymbol{y} \in R^I : y_i + y_j \geq 6\}, \\
&= 1 && \text{for} \quad \boldsymbol{x} \notin \mathcal{U}(K) \\
& && \qquad \text{and} \quad \boldsymbol{x} \in \{\boldsymbol{y} \in R^I : y_i + y_j = 5\}, \\
&= 6 - (x_i + y_j) && \text{for} \quad x \notin \mathcal{U}(K) \\
& && \qquad \text{and} \quad \boldsymbol{x} \in \{\boldsymbol{y} \in R^I : 5 < y_i + y_j < 6\}.
\end{aligned}
$$

□

Example 10.8 Let us consider a coalition game $(I, \mathcal{V})$ and a coalition $K = \{i, j\} \subset I$ already treated in the previous example. Analogously to Example 10.7 we also denote

$$
\begin{aligned}
\mathcal{U}(K) \;=\; & \{\boldsymbol{x} \in R^I : x_i > 5\} \cup \{\boldsymbol{x} \in R^I : x_j > 5\} \cup \\
& \cup \{\boldsymbol{x} \in R^I : x_i = 5 \text{ and } x_j < 0\} \cup \{\boldsymbol{x} \in R^I : x_j = 5 \text{ and } x_i < 0\}.
\end{aligned}
$$

The fuzzy extension $\mathcal{W}(K)$ of $\mathcal{V}(K)$ is defined by

$$
\begin{aligned}
\kappa_K(\boldsymbol{x}) &= 0 && \text{for } \boldsymbol{x} \in \mathcal{U}(K) \cup \{\boldsymbol{y} \in R^I : y_i + y_j > 6\}, \\
&= 1/2 && \text{for } \boldsymbol{x} \notin \mathcal{U}(K) \text{ and } \boldsymbol{x} \in \{\boldsymbol{y} \in R^I : 5 < y_i + y_j \leq 6\}, \\
&= 1 && \text{for } \boldsymbol{x} \notin \mathcal{U}(K) \text{ and } \boldsymbol{x} \in \{\boldsymbol{y} \in R^I : y_i + y_j \leq 5\}.
\end{aligned}
$$

Then, we can easily see that

$$
\begin{aligned}
\kappa_K^*(\boldsymbol{x}) &= 0 && \text{for } \boldsymbol{x} \notin \mathcal{U}(K) \text{ and } \boldsymbol{x} \in \{\boldsymbol{y} \in R^I : y_i + y_j < 5\}, \\
&= 1 && \text{for } \boldsymbol{x} \notin \mathcal{U}(K) \text{ and } \boldsymbol{x} \in \{\boldsymbol{y} \in R^I : y_i + y_j = 5\} \cup \\
& && \qquad \cup \{\boldsymbol{y} \in R^I : y_i + y_j \geq 6\}, \\
&= 1/2 && \text{for } \boldsymbol{x} \notin \mathcal{U}(K) \text{ and } \boldsymbol{x} \in \{\boldsymbol{y} \in R^I : 5 < y_i + y_j < 6\},
\end{aligned}
$$

and, consequently,

$$
\begin{aligned}
\rho_K(\boldsymbol{x}) &= 0 && \text{for } \boldsymbol{x} \in \mathcal{U}(K) \cup \{\boldsymbol{y} \in R^I : y_i + y_j < 5\} \cup \\
& && \qquad \cup \{\boldsymbol{y} \in R^I : y_i + y_j > 6\}, \\
&= 1/2 && \text{for } \boldsymbol{x} \notin \mathcal{U}(K) \text{ and } \boldsymbol{x} \in \{\boldsymbol{y} \in R^I : 5 < y_i + y_j \leq 6\}, \\
&= 1 && \text{for } \boldsymbol{x} \notin \mathcal{U}(K) \text{ and } \boldsymbol{x} \in \{\boldsymbol{y} \in R^I : y_i + y_j - 5\}.
\end{aligned}
$$

□

11. Fuzzy and Imputational Additivity–Like Relations

Intuitively, we can expect that the superadditivity, subadditivity or additivity in fuzzy coalition games without side-payments are to be defined analogously to their deterministic counterparts and, in this sense, they become extensions of deterministic concepts. If we do so and formulate close analogies of the concepts formulated in Subsection 3.2, we can see that the additivity-like properties in fuzzy coalition games appear to be deterministic what rather contradicts our heuristic assumption that important properties of fuzzy games would also be fuzzy. To overcome this gap we suggest a modification of this approach and to formulate it as a fuzzy property.

Generally, while addivity-like properties in fuzzy coalition games with side-payments are based on an ordering relation over fuzzy quantities, in fuzzy coalition games without side-payments they follow from inclusion relations between sets of (fuzzy) admissible imputations.

These two qualitatively different approaches lead to different results mostly regarding the relation between superadditivity (subadditivity, additivity) of a deterministic game and its fuzzy extension.

11.1 Imputational Superadditivity

Let us consider a coalition game without side-payments $(I, \mathcal{V})$ and its fuzzy extension $(I, \mathcal{W})$. Analogously to the deterministic case we say that $(I, \mathcal{W})$ is *imputationally superadditive* iff for any pair of disjoint coalitions $K, L \subset I$, $K \cap L = \emptyset$, there holds the inclusion

$$\mathcal{W}(K \cup L) \supset \mathcal{W}(K) \cap \mathcal{W}(L) \tag{11.1}$$

which means that

$$\kappa_{K \cup L}(\boldsymbol{x}) \geq \min(\kappa_K(\boldsymbol{x}), \kappa_L(\boldsymbol{x})) \quad \text{for any} \quad \boldsymbol{x} \in R^I \tag{11.2}$$

holds.

Due to this definition, the imputational superadditivity of $(I, \mathcal{W})$ is a deterministic property – relations (11.1) and (11.2) are either true or false. As mentioned above, this fact rather contradicts our assumption that characteristic properties of fuzzy coalition games are to be fuzzy. However, even the simple "deterministic" imputational superadditivity introduced above implies some interesting properties. Some of them are mentioned in [36] and [34].

Let us consider a fuzzy extension $(I, \mathcal{W})$ of a game $(I, \mathcal{V})$ such that for all $K \subset I$, $\kappa_K(\boldsymbol{x}) = 1$ iff $\boldsymbol{x} \in \mathcal{V}(K)$. If $(I, \mathcal{W})$ is imputationally superadditive, then $(I, \mathcal{V})$ is superadditive in the sense of (3.19).

Analogously to the games with side-payments, even for any coalition game without side-payments $(I, \mathcal{V})$ there always exists its fuzzy extension $(I, \mathcal{W})$ which is imputationally superadditive. Due to [36], this statement can be modified. Namely, for any superadditive game $(I, \mathcal{V})$ there exists its fuzzy extension $(I, \mathcal{W})$ such that for all $K \subset I$, $\kappa_K(\boldsymbol{x}) = 1$ if and only if $\boldsymbol{x} \in \mathcal{V}(K)$, which is imputationally s,uperadditive.

If $(I, \mathcal{W})$ and $(I, \mathcal{W}')$ are imputationally superadditive fuzzy extensions of a game $(I, \mathcal{V})$, then the fuzzy coalition game $(I, [\mathcal{W} \cap \mathcal{W}'])$, where – for any $K \subset I$ – $[\mathcal{W} \cap \mathcal{W}'](K) = \mathcal{W}(K) \cap \mathcal{W}'(K)$ is imputationally superadditive.

If we want to respect the heuristic principle due to which principal properties of fuzzy coalition games are to be fuzzy, then we would modify the "deterministic" imputational superadditivity (11.1) by another one, enabling to respect the structured validity of vague superadditivity. Quite a natural approach to this task was suggested in [34] and [36].

11.2 Fuzzy Superadditivity

Attempting to implement vagueness into the concept of superadditivity in fuzzy coalition games without side-payments, we use the notion of α-cuts of fuzzy sets.

If $(I, \mathcal{W})$ is a fuzzy extension of $(I, \mathcal{V})$, if $K \subset I$ and $\alpha \in [0, 1]$, then an *α-cut of the fuzzy set* $\mathcal{W}(K)$ is fuzzy set $\mathcal{W}_\alpha(K)$ with the membership function $\kappa_K^{(\alpha)} : R^I \to [0, \alpha]$ such that for $\boldsymbol{x} \in R^I$

$$\kappa_K^{(\alpha)}(\boldsymbol{x}) = \min(\kappa_K(\boldsymbol{x}), \alpha). \tag{11.3}$$

The pair $(I, \mathcal{W}_\alpha)$ will be called an *α-reduction* of the game $(I, \mathcal{W})$.

Now, we define an auxiliary relation, termed a *fuzzy inclusion with possibility* $\alpha \in (0, 1]$ and denoted by $\supset_\alpha$, given for any pair of disjoint coalitions as

$$\mathcal{W}(K \cup L) \supset_\alpha \mathcal{W}(K) \cap \mathcal{W}(L) \tag{11.4}$$

iff

$$\mathcal{W}_\alpha(K \cup L) \supset \mathcal{W}_\alpha(K) \cap \mathcal{W}_\alpha(L)$$

that is

$$\kappa_{K\cup L}^{(\alpha)}(\boldsymbol{x}) \geq \min(\kappa_K^{(\alpha)}(\boldsymbol{x}), \kappa_L^{(\alpha)}(\boldsymbol{x})) \quad \text{for all} \quad \boldsymbol{x} \in R^I.$$

Evidently, if $\mathcal{W}(K \cup L) \supset_\alpha \mathcal{W}(K) \cap \mathcal{W}(L)$, then also $\mathcal{W}(K \cup L) \supset_\beta \mathcal{W}(K) \cap \mathcal{W}(L)$, for all $\beta \in [0, \alpha]$.

It is obvious that the *fuzzy superadditivity* can be defined as a fuzzy property of $(I, \mathcal{W})$ which is valid with possibility

$$\nu_{\text{super}}(I, \mathcal{W}) = \sup \Big(\alpha \in [0,1] : \forall K, L \subset I \text{ such that } K \cap L = \emptyset,$$

$$\text{the inclusion } \mathcal{W}(K \cup L) \supset_\alpha \mathcal{W}(K) \cap \mathcal{W}(L) \text{ holds } \Big). \quad (11.5)$$

It is easy to see that if $\nu_{\text{super}}(I, \mathcal{W}) \neq \alpha$ for all $\alpha \in (0, 1]$, then $\nu_{\text{super}}(I, \mathcal{W}) = 0$ because, for all $K, L \subset I$, $K \cap L = \emptyset$, the inclusion $\mathcal{W}_0(K \cup L) \supset \mathcal{W}_0(K) \cap \mathcal{W}_0(L)$, where $\mathcal{W}_0(K \cup L) = \mathcal{W}_0(K) = \mathcal{W}_0(L) = R^I$, holds.

In the usual fuzzy set theoretical interpretation of (11.5), $\nu_{\text{super}}(I, \mathcal{W})$ is the possibility that the α-cuts of $\mathcal{W}(K \cup L)$, $\mathcal{W}(K)$, $\mathcal{W}(L)$ fulfill the defining relation of superadditivity for all pairs of disjoint coalitions and, moreover, that this relation is not valid with any higher possibility.

Equality (11.5) implies that $\nu_{\text{super}}(I, \mathcal{W}) = 1$ iff $(I, \mathcal{W})$ is imputationally superadditive. Using the properties presented in the previous subsection, we can easily see that if $\nu_{\text{super}}(I, \mathcal{W}) = 1$ and if $\kappa_K(\boldsymbol{x}) = 1$ iff $\boldsymbol{x} \in \mathcal{V}(K)$, then the original deterministic game $(I, \mathcal{V})$ is superdditive.

It is also possible to formulate an analogy of a statement formulated for imputational superadditivity. If $\beta \in (0, 1]$ and if $(I, \mathcal{V})$ is a coalition game, then there always exists its fuzzy extension $(I, \mathcal{W})$ such that $\nu_{\text{super}}(I, \mathcal{W}) = \beta$.

The previous statements immediately follow from the results derived in [36]. They illustrate, from a general point of view, the fact mentioned already in the introductory paragraphs of this chapter.

Namely, there exist some relations between the superadditivities of the original deterministic games without side-payments and their fuzzy extensions. But they are of a different type than those found in the case of coalition games with side-payments. They are also, in a certain sense, weaker and also the concept of superadditivity (especially of the imputational one) is less influenced by vagueness which is present in the fuzzy extension of the game.

This difference between the fuzzy extensions of games with and without side-payments follows from the difference between their mathematical models. The processing of fuzzy quantities differs from the processing of fuzzy subsets of R^I and the ordering relation over fuzzy quantities essentially differs from the inclusion relation between fuzzy sets.

A certain interconnection between the superadditivities in deterministic coalition games with and without side-payments shown, e.g., in [20] is much weaker if the fuzzy extensions are considered. More about that interconnection is mentioned in Chapter 14.

The concepts of imputational and fuzzy superadditivity can be illustrated by the following examples. They partly use our experience with dominance, in particular formulated in Example 10.1 and Example 10.7.

Example 11.1 Let us consider a 2-players coalition game $(I, \mathcal{V})$ with $I = \{1, 2\}$, and

$$\begin{aligned}\mathcal{V}(\{i\}) &= \{\boldsymbol{x} \in R^2 : x_i \le 2\}, \quad i = 1, 2,\\ \mathcal{V}(I) &= \{\boldsymbol{x} \in R^2 : x_1 + x_2 \le 5\} \cap \{\boldsymbol{x} \in R^2 : x_1 \le 5\} \cap \{\boldsymbol{x} \in R^2 : x_2 \le 5\}.\end{aligned}$$

We will study its fuzzy extension $(I, \mathcal{W})$, where

$$\begin{aligned}\kappa_I(\boldsymbol{x}) &= 1 && \text{for } \boldsymbol{x} \in \{\boldsymbol{y} \in R^2 : y_1 + y_2 \le 5\} \cap\\ &&& \cap \{\boldsymbol{y} \in R^2 : y_1 \le 5\} \cap\\ &&& \cap \{\boldsymbol{y} \in R^2 : y_1 \le 5\} \cap \{\boldsymbol{y} \in R^2 : y_2 \le 5\},\\ &&& \text{i.e. } \boldsymbol{x} \in \mathcal{V}(I),\\ &= 6 - (x_1 + x_2) && \text{for } \boldsymbol{x} \in \{\boldsymbol{y} \in R^2 : 5 < y_1 + y_2 \le 6\} \cap\\ &&& \cap \{\boldsymbol{y} \in R^2 : y_1 \le 5\} \cap \cap \{\boldsymbol{y} \in R^2 : y_2 \le 5\},\\ &= 0 && \text{for } x_1 + x_2 \ge 6 \text{ or } x_1 > 5 \text{ or } x_2 > 5.\end{aligned}$$

$$\begin{aligned}\kappa_{\{i\}}(\boldsymbol{x}) &= 1 && \text{for } x_i \le 2,\\ &= 3 - x_i && \text{for } x_i \in (2, 3),\\ &= 0 && \text{for } x_i \ge 3.\end{aligned}$$

It is easy to see that for, all $\boldsymbol{x} \in R^2$, we have

$$\kappa_I(\boldsymbol{x}) \ge \min\left(\kappa_{\{1\}}(\boldsymbol{x}), \kappa_{\{2\}}(\boldsymbol{x})\right)$$

and, consequently, $(I, \mathcal{W})$ is imputationally superadditive or fuzzy superadditive with possibility $\nu_{\text{super}}(I, \mathcal{W}) = 1$.

□

Example 11.2 Let us consider the coalition game $(I, \mathcal{V})$ already discussed in the previous example, and its fuzzy extension $(I, \mathcal{W})$ such that for any $\boldsymbol{x} \in R^2$, there holds

for $i = 1, 2$

$$\begin{aligned}\kappa_{\{i\}}(\boldsymbol{x}) &= 1 && \text{for } \ x_1 \le 3,\\ &= (15 - 4x_i)/3 && \text{for } \ 3 \le x_i \le 15/4,\\ &= 0 && \text{for } \ x_i \ge 15/4.\end{aligned}$$

$$\begin{aligned}\kappa_I(\boldsymbol{x}) &= 1 && \text{for } \boldsymbol{x} \in \{\boldsymbol{y} \in R^2 : y_1 + y_2 \le 5\} \cap\\ &&& \cap \{\boldsymbol{y} \in R^2 : y_1 \le 5\} \cap \{\boldsymbol{y} \in R^2 : y_2 \le 5\},\\ &= (8 - (x_1 + x_2))/3 && \text{for } \boldsymbol{x} \in \{\boldsymbol{y} \in R^2 : 5 \le y_1 + y_2 \le 8\} \cap\\ &&& \cap \{\boldsymbol{y} \in R^2 : y_1 \le 5\} \cap \{\boldsymbol{y} \in R^2 : y_2 \le 5\},\\ &= 0 && \text{for } x_1 > 5 \text{ or } x_2 > 5 \text{ or } x_1 + x_2 \ge 8.\end{aligned}$$

This fuzzy coalition game $(I, \mathcal{W})$ is not imputationally superadditive as, e.g., for $\boldsymbol{x} = (3,3)$, we have

$$1 = \min\left(\kappa_{\{1\}}(\boldsymbol{x}), \kappa_{\{2\}}(\{\boldsymbol{x}\})\right) > \kappa_I(\boldsymbol{x}) = 2/3.$$

However, it is not difficult to verify that for every $\alpha \leq 1/3$

$$\kappa_I^{(\alpha)} \geq \min\left(\kappa_{\{1\}}^{(\alpha)}(\boldsymbol{x}), \kappa_{\{2\}}^{(\alpha)}(\{\boldsymbol{x}\})\right) \quad \text{for all } \boldsymbol{x} \in R^2$$

and, consequently, $\nu_{\text{super}}(I, \mathcal{W}) = 1/3$.

□

In fuzzy extensions of coalition games without side-payments the superadditivity of the deterministic game $(I, \mathcal{V})$ does not imply the (imputational or fuzzy) superadditivity.

In this property these games differ from the games with side-payments. It can be illustrated by the following example.

Example 11.3 Let us consider the coalition game $(I, \mathcal{V})$ processed in the previous two examples, and its fuzzy extension $(I, \mathcal{W})$ with

$$\begin{aligned} \kappa_I(\boldsymbol{x}) &= 1 \quad && \text{for } \boldsymbol{x} \in \mathcal{V}(I), \\ &= 0 && \text{for } \boldsymbol{x} \notin \mathcal{V}(I), \end{aligned}$$

$$\begin{aligned} \kappa_{\{i\}}(\boldsymbol{x}) &= 1 \quad && \text{for } x_1 \leq 2, \\ &= 3 - x_i && \text{for } x_i \in (2,3) \\ &= 0 && \text{for } x_i \geq 3 \end{aligned}$$

for all $\boldsymbol{x} \in R^2$.

Then, it is easy to see that

$$\begin{aligned} \min\left(\kappa_{\{1\}}(\boldsymbol{x}), \kappa_{\{2\}}(\boldsymbol{x})\right) &= 1 \text{ for } \boldsymbol{x} \in \{\boldsymbol{y} \in R^2 : y_1 \leq 2\} \cap \\ &\quad \cap \{\boldsymbol{y} \in R^2 : y_2 \leq 2\}, \\ &= 0 \text{ for } \boldsymbol{x} \in R^2 \text{ such that} \\ &\quad x_1 \geq 3 \text{ or } x_2 \geq 3, \\ &= 3 - x_1 \text{ for } \boldsymbol{x} \in \{\boldsymbol{y} \in R^2 : 2 < y_1 < 3, y_2 \leq y_1\} \\ &= 3 - x_2 \text{ for } \boldsymbol{x} \in \{\boldsymbol{y} \in R^2 : 2 < y_1 < 3, y_1 \leq y_2\}. \end{aligned}$$

Therefore, obviously,

$$\min\left(\kappa_{\{1\}}(\boldsymbol{x}), \kappa_{\{2\}}(\boldsymbol{x})\right) > \kappa_I(\boldsymbol{x}) = 0$$

for $\boldsymbol{x} \in R^2$ such that $x_1 + x_2 > 5$, $x_1 < 3$, $x_2 < 3$, i.e. for $\boldsymbol{x}$ from the inner part of the triangle determined by vertices $(3,2), (2,3), (3,3)$. Hence,

$(I, \mathcal{W})$ is neither imputationally superadditive nor fuzzy superadditive with any positive possibility.

□

11.3 Imputational Subadditivity and Additivity

The subadditivity of fuzzy extensions of coalition games without side-payments can be approached analogously to their superadditivity. It is evident that the analogy of results is limited by specific properties of the fuzzy superoptima $\mathcal{W}^*(K)$ for $K \subset I$. First, the fuzzy sets $\mathcal{W}^*(K)$ need not be monotone in the sense that for $\boldsymbol{x}, \boldsymbol{y} \in R^I$, $\boldsymbol{x} \operatorname{dom}_K \boldsymbol{y}$ the inequality $\kappa_K(\boldsymbol{x}) \geq \kappa_K(\boldsymbol{y})$ need not be generally fulfilled. Second, the fuzzy sets $\mathcal{W}^*(K)$ are, in the fuzzy set theoretic sense, subsets of $\mathcal{V}^*(K)$ (cf. Chapter 10). On the contrary, the sets $\mathcal{V}(K)$ are fuzzy subsets of $\mathcal{W}(K)$. Third, there exist a significant class of coalitions, namely those for which $\kappa_K(\boldsymbol{x})$ is strictly decreasing in all coordinates for $\boldsymbol{x}$ from the set

$$\{\boldsymbol{x} \in R^I : 0 < \kappa_K(\boldsymbol{x}) < 1\},$$

where $\mathcal{W}^*(K) = \mathcal{V}^*(K)$ in the fuzzy set theoretical sense, i.e. the characteristic function $\kappa_K^* : R^I \to \{0, 1\}$ represents the characteristic function of $\mathcal{V}^*(K)$.

These properties, already presented in Chapter 10, limit the relations between the subadditivities of the original deterministic games and their fuzzy extensions.

First of all, we formulate a definition, quite analogous to the deterministic one. If $(I, \mathcal{V})$ is a coalition game without side-payments and $(I, \mathcal{W})$ is its fuzzy extension, then we say that $(I, \mathcal{W})$ is *imputationally subadditive* iff for any pair of disjoint coalitions $K, L \subset I$, $K \cap L = \emptyset$, there holds

$$\mathcal{W}^*(K \cup L) \supset \mathcal{W}^*(K) \cap \mathcal{W}^*(L), \tag{11.6}$$

which means

$$\kappa_{K \cup L}^*(\boldsymbol{x}) \geq \min(\kappa_K^*(\boldsymbol{x}), \kappa_L^*(\boldsymbol{x})) \quad \text{for all} \quad \boldsymbol{x} \in R^I. \tag{11.7}$$

The properties of superoptima $\mathcal{W}^*(K)$, $K \subset I$, mentioned in Chapter 10, imply the following conclusions.

If for any $K \subset I$, $\{\boldsymbol{x} \in R^I : \kappa_K(\boldsymbol{x}) = 1\} = \mathcal{V}(K)$, then also

$$\mathcal{V}^*(K) = \{\boldsymbol{x} \in R^I : \kappa_K^*(\boldsymbol{x}) > 0\}$$

and, consequently, if for a pair of disjoint coalitions, $K, L \subset I$ (11.7) is true, then also

$$\mathcal{V}^*(K \cup L) \supset \mathcal{V}^*(K) \cap \mathcal{V}^*(L).$$

This means that in such a case the subadditivity of $(I, \mathcal{W})$ implies the subadditivity of $(I, \mathcal{V})$. Also this statement can be extended analogously to a similar property of superoptima. If $(I, \mathcal{W})$ is a fuzzy extension of $(I, \mathcal{V})$ and if for any $K \subset I$ the set $\overline{\mathcal{V}}(K) \subset R^I$ is such that

$$\overline{\mathcal{V}}(K) = \{\boldsymbol{x} \in R^I : \kappa_K(\boldsymbol{x}) = 1\} \tag{11.8}$$

then the pair $(I, \overline{\mathcal{V}})$ is also a coalition game and it is subadditive if $(I, \mathcal{W})$ is subadditive. It is also easy to show (see [36]) that for any game $(I, \mathcal{V})$ there exists its fuzzy extension $(I, \mathcal{W})$ which is subadditive.

Of course, if for any $K \subset I$, $\{\boldsymbol{x} \in R^I : \kappa_K(\boldsymbol{x}) = 1\} = \mathcal{V}(K)$ and $\kappa_K(\boldsymbol{x})$ is strictly decreasing for all $\boldsymbol{x}$ from the set $\{\boldsymbol{x} \in R^I : \kappa_K(\boldsymbol{x}) \in (0,1)\}$ in the sense presented in Chapter 10, then $(I, \mathcal{W})$ is subadditive if $(I, \mathcal{V})$ is subadditive since

$$\mathcal{V}^*(K) = \{\boldsymbol{x} \in R^I : \kappa_K^*(\boldsymbol{x}) > 0\} \quad \text{and} \quad \mathcal{W}^*(K) = \mathcal{V}^*(K)$$

for any coalition $K \subset I$.

Analogously to the deterministic game, we say that a fuzzy extension $(I, \mathcal{W})$ of a coalition game is *imputationally additive* iff it is imputationally superadditive and imputationally subadditive.

This means that for any pair of disjoint coalitions $K, L \subset I$, $K \cap L = \emptyset$, we have

$$\mathcal{W}(K \cup L) \supset \mathcal{W}(K \cap \mathcal{W}(L)) \quad \text{and} \quad \mathcal{W}^*(K \cup L) \supset \mathcal{W}^*(K) \cap \mathcal{W}^*(L),$$

which means that

$$\kappa_{K \cup L}(\boldsymbol{x}) \geq \min(\kappa_K(\boldsymbol{x}), \kappa_L(\boldsymbol{x})) \quad \text{and} \quad \kappa^*_{K \cup L}(\boldsymbol{x}) \geq \min(\kappa^*_K(\boldsymbol{x}), \kappa^*_L(\boldsymbol{x}))$$

for all $\boldsymbol{x} \in R^I$.

Even the imputational additivity is, similarly to the superadditivity and subadditivity, an essentially deterministic phenomenon. Its following properties are immediate consequences of their analogues that are valid for the superadditivity and subadditivity.

For any coalition game $(I, \mathcal{V})$ there exists its fuzzy extension $(I, \mathcal{W})$ which is additive. If for any $K \subset I$, $\{\boldsymbol{x} \in R^I : \kappa_K(\boldsymbol{x}) = 1\} = \mathcal{V}(K)$, then the additivity of the fuzzy extension $(I, \mathcal{W})$ implies the additivity of the original deterministic game $(I, \mathcal{V})$.

More generally, if $(I, \mathcal{W})$ is a fuzzy extension of $(I, \mathcal{V})$ and if for any $K \subset I$ we denote $\overline{\mathcal{V}}(K)$ due to (11.8), then the additivity of $(I, \mathcal{W})$ implies the additivity of the deterministic coalition game $(I, \overline{\mathcal{V}})$.

11.4 Fuzzy Subadditivity and Additivity

Similarly as the imputational superadditivity, also the (imputational) subadditivity and additivity are in their nature crisp properties offering only the yes/no type validity. If we want to include a scale of possibilities into the subadditivity and additivity concepts in the considered type of fuzzy coalition games, then we can proceed analogously to the superadditivity case and use the method of α-cuts of fuzzy sets and their membership functions.

We use the definition (11.3) of α-cuts $\mathcal{W}_\alpha^*(K)$ of $\mathcal{W}^*(K)$ for $K \subset I$, $\alpha \in [0,1]$ and the definition of fuzzy inclusion (11.4).

Then the *fuzzy subadditivity* can be defined as a fuzzy property of $(I, \mathcal{W})$ which is valid with the possibility

$$\nu_{\text{sub}}(I, \mathcal{W}) = \sup\Big(\alpha \in [0,1] : \forall\, K, L \subset I, \text{ such that } K \cap L = \emptyset$$
$$\text{the inclusion } \mathcal{W}^*(K \cup L) \supset_\alpha \mathcal{W}^*(K) \cap \mathcal{W}^*(L) \text{ holds}\Big). \quad (11.9)$$

It is evident that, similarly to the fuzzy superadditivity, if $\nu_{\text{sub}}(I, \mathcal{W}) \neq \alpha$ for all $\alpha \in (0,1]$, then (11.9) implies $\nu_{\text{sub}}(I, \mathcal{W}) = 0$ since for all $K \subset I$, $\mathcal{W}_0^* = R^I$.

Let for all $K \subset I$ the membership function $\kappa_K(\boldsymbol{x})$ be strictly decreasing for all $\boldsymbol{x} \in R^I$ such that $\kappa_K(\boldsymbol{x}) \in (0,1)$ in the sense that $\kappa_K(\boldsymbol{x}) < \kappa_K(\boldsymbol{y})$ if $\boldsymbol{x}\, \mathrm{dom}_K\, \boldsymbol{y}$, and $\kappa_K(\boldsymbol{x}) \in [0,1)$, $\kappa_K(\boldsymbol{y}) \in (0,1]$. Then, obviously, the membership functions κ_K^* of $\mathcal{W}^*(K)$ take on two values only, namely 0 and 1 (cf. Chapter 10), where $\kappa_K^*(\boldsymbol{x}) = 1$ iff $\kappa_K(\boldsymbol{x}) < 1$.

Similarly, the membership functions $\kappa_K^{*(\alpha)}$ of $\mathcal{W}_\alpha^*(K)$, $\alpha \in [0,1]$, take on only values 0 and α with $\kappa_K^{*(\alpha)}(\boldsymbol{x}) = \alpha$ iff $\kappa_K^*(\boldsymbol{x}) = 1$ and $\kappa_K^{*(\alpha)}(\boldsymbol{x}) = 0$ for $\boldsymbol{x} \notin \mathcal{V}^*(K)$.

Moreover, in such game, the possibility $\nu_{\text{sub}}(I, \mathcal{W})$ is either equal to 1 if $(I, \mathcal{W})$ is imputationally subadditive, or it is equal to 0 in the opposite case. In other words, for the considered type of fuzzy games with strictly decreasing membership functions, for any disjoint $K, L \subset I$ and any $\alpha \in (0,1]$, there holds

$$\mathcal{W}^*(K \cup L) \supset_\alpha \mathcal{W}^*(K) \cap \mathcal{W}^*(L) \quad \text{iff} \quad \mathcal{W}^*(K \cup L) \supset \mathcal{W}^*(K) \cap \mathcal{W}^*(L).$$

Generally, if $(I, \mathcal{W})$ is a fuzzy extension of $(I, \mathcal{V})$, if $\alpha, \beta \in [0,1]$, $\alpha \geq \beta$ and if $K, L \subset I$, $K \cap L = \emptyset$, then

$$\mathcal{W}_\alpha^*(K \cup L) \supset \mathcal{W}_\alpha^*(K) \cap \mathcal{W}_\alpha^*(L) \quad \text{implies} \quad \mathcal{W}_\beta^*(K \cup L) \supset \mathcal{W}_\beta^*(K) \cap \mathcal{W}_\beta^*(L).$$

Moreover, $\nu_{\text{sub}}(I, \mathcal{W}) = 1$ if and only if $(I, \mathcal{W})$ is imputationally subadditive. If $\nu_{\text{sub}}(I, \mathcal{W}) = 1$ and $\kappa_K(\boldsymbol{x}) = 1$ iff $\boldsymbol{x} \in \mathcal{V}(K)$, then the original deterministic game $(I, \mathcal{V})$ is subadditive.

It is also easy to verify that for every $\beta \in (0,1]$ and every game $(I, \mathcal{V})$, there exists its fuzzy extension $(I, \mathcal{W})$ such that $\nu_{\text{sub}}(I, \mathcal{W}) = \beta$.

Having defined the concepts of fuzzy subperadditivity and fuzzy subadditivity we can, respecting the approach used in the previous sections, define the fuzzy additivity as a fuzzy logical conjunction of the (fuzzy) superadditivity and subadditivity.

Exactly formulated, we say that a fuzzy extension $(I, \mathcal{W})$ of a coalition game $(I, \mathcal{V})$ is *fuzzy additive* iff it is fuzzy superadditive and fuzzy subadditive. The possibility that $(I, \mathcal{W})$ is fuzzy additive is

$$\nu_{\text{addit}}(I, \mathcal{W}) = \min(\nu_{\text{super}}(I, \mathcal{W}), \nu_{\text{sub}}(I, \mathcal{W})). \tag{11.10}$$

The properties of the fuzzy superadditivity and subadditivity immediately imply some statements regarding the fuzzy additivity. Namely, for every coalition game $(I, \mathcal{V})$ and for every $\beta \in (0, 1]$ there exists its fuzzy extension $(I, \mathcal{W})$ which is additive with possibility $\nu_{\text{addit}}(I, \mathcal{W}) = \beta$, $(I, \mathcal{W})$ is imputationally additive iff it is fuzzy additive with $\nu_{\text{addit}}(I, \mathcal{W}) = 1$.

If $(I, \mathcal{W})$ is a fuzzy extension of $(I, \mathcal{V})$ such that for all $K \subset I$, $\kappa_K(\boldsymbol{x}) = 1$ iff $\boldsymbol{x} \in \mathcal{V}(K)$ and if $(I, \mathcal{W})$ is fuzzy additive with possibility $\nu_{\text{addit}}(I, \mathcal{W}) = 1$, then $(I, \mathcal{V})$ is additive.

Also the above properties of (imputational and fuzzy) subadditivity and additivity can be illustrated by simple examples. In the ones presented below we partly use our experience following from Example 9.3 and Example 10.3.

Example 11.4 Let us consider a fuzzy coalition game $(I, \mathcal{V})$ with 2 players, $I = \{1, 2\}$, and

$$\begin{aligned} \mathcal{V}(I) &= \{\boldsymbol{x} \in R^2 : x_1 \cdot x_2 \le 1\}, \\ \mathcal{V}(\{i\}) &= \{\boldsymbol{x} \in R^2 : x_1 \le 1\}, \quad i = 1, 2. \end{aligned}$$

This game is subadditive as

$$\mathcal{V}^*(I) = \{\boldsymbol{x} \in R^2 : x_1 \cdot x_2 \ge 1\}, \quad \mathcal{V}^*(\{i\}) = \{\boldsymbol{x} \in R^2 : x_1 \ge 1\}, \quad i = 1, 2.$$

Let us consider its fuzzy extension $(I, \mathcal{W})$ such that

$$\begin{aligned} \kappa_I(\boldsymbol{x}) &= 1 && \text{for} \quad \boldsymbol{x} \in \{\boldsymbol{y} \in R^2 : y_1 \cdot y_2 \le 1\}, \\ &= 0 && \text{for} \quad \boldsymbol{x} \in \{\boldsymbol{y} \in R^2 : (y_1 - 1) \cdot (y_2 - 1) \ge 1\}, \\ &= 1 - \lambda && \text{for} \quad \boldsymbol{x} \in \{\boldsymbol{y} \in R^2 : (y_1 - \lambda) \cdot (y_2 - \lambda) = 1\}, \ 0 < \lambda < 1, \end{aligned}$$

$$\begin{aligned} \kappa_{\{i\}} &= 1 && \text{for} \quad x_i \le 1, \\ &= 0 && \text{for} \quad x_i > 2, \\ &= 1/2 && \text{for} \quad 1 < x_1 \le 2. \end{aligned}$$

Then, it is not difficult to verify that $\mathcal{W}^*(I) = \mathcal{V}^*(I)$, i.e.

$$\begin{aligned} \kappa_I^*(\boldsymbol{x}) &= 0 \quad \text{for} \quad x_1 \cdot x_2 < 1, \\ &= 1 \quad \text{for} \quad x_1 \cdot x_2 \ge 1. \end{aligned}$$

Similarly,

$$\begin{aligned}\kappa^*_{\{i\}}(\boldsymbol{x}) &= 0 && \text{for} && x_1 < 1,\\ &= 1 && \text{for} && x_i = 1,\\ &= 1/2 && \text{for} && 1 < x_1 < 2,\\ &= 1 && \text{for} && x_1 \geq 2.\end{aligned}$$

Then, evidently, for all $\boldsymbol{x} \in R^2$ we have

$$\kappa^*_I(\boldsymbol{x}) \geq \min\left(\kappa^*_{\{1\}}(\boldsymbol{x}), \kappa^*_{\{2\}}(\boldsymbol{x})\right)$$

and $(I, \mathcal{W})$ is imputationally subadditive as well as fuzzy subadditive with possibility $\nu_{\text{sub}}(I, \mathcal{W}) = 1$.

□

Example 11.5 Let us consider the coalition game $(I, \mathcal{V})$ which was discussed in the previous example and its fuzzy extension $(I, \mathcal{W})$ such that for all $\boldsymbol{x} \in R^2$, $i = 1, 2$,

$$\begin{aligned}\kappa_{\{i\}} &= 1 && \text{for} && x_1 \leq 1,\\ &= 0 && \text{for} && x_1 > 2,\\ &= 1/2 && \text{for} && 1 < x_i \leq 2,\end{aligned}$$

(these fuzzy sets $\mathcal{W}(\{1\}), \mathcal{W}(\{2\})$ were considered in Example 11.4 too), and

$$\begin{aligned}\kappa_I(\boldsymbol{x}) &= 1 && \text{for} && x_1 \cdot x_2 \leq 1,\\ &= 0 && \text{for} && (x_1 - 1) \cdot (x_2 - 1) \geq 1,\\ &= 1/2 && \text{for} && (x_1 - \lambda) \cdot (x_2 - \lambda) = 1, \ \lambda \in (0, 1).\end{aligned}$$

Then, the membership functions $\kappa^*_{\{1\}}$ and $\kappa^*_{\{2\}}$ were already computed in Example 11.4, and for $\boldsymbol{x} \in R^2$

$$\begin{aligned}\kappa^*_I(\boldsymbol{x}) &= 0 && \text{for} && x_1 \cdot x_2 < 1,\\ &= 1 && \text{for} && x_1 \cdot x_2 = 1,\\ &= 1/2 && \text{for} && (x_1 - \lambda) \cdot (x_2 - \lambda) = 1, \quad \lambda \in (0, 1),\\ &= 1 && \text{for} && (x_1 - 1) \cdot (x_2 - 1) \geq 1.\end{aligned}$$

We can see that $(I, \mathcal{W})$ is not imputationally subadditive as, e.g., for $\boldsymbol{x} = (1, 2)$,

$$\min\left(\kappa^*_{\{1\}}(\boldsymbol{x}), \kappa^*_{\{2\}}(\boldsymbol{x})\right) = 1 > \kappa^*_I(\boldsymbol{x}) = 1/2.$$

But, it is also not difficult to see that for all $\alpha \leq 1/2$ and all $\boldsymbol{x} \in R^I$

$$\begin{aligned}\kappa^{*(\alpha)}_I(\boldsymbol{x}) &= 0 && \text{for} && x_1 \cdot x_2 < 1\\ &= 1/2 && \text{for} && (x_1 - \lambda) \cdot (x_2 - \lambda) = 1, \ \lambda \in [0, 1),\\ &= 1/2 && \text{for} && (x_1 - \lambda) \cdot (x_2 - \lambda) > 1,\end{aligned}$$

$$\begin{aligned}\kappa^*_{\{i\}}(\boldsymbol{x}) &= 0 && \text{for} && x_i < 1,\\ &= 1/2 && \text{for} && x_1 \geq 1.\end{aligned}$$

It means that $\mathcal{W}^{*(\alpha)}(I) \supset \mathcal{W}^{*(\alpha)}(\{1\}) \cap \mathcal{W}^{*(\alpha)}(\{2\})$, and, consequently, $(I, \mathcal{W})$ is fuzzy subadditive with possibility $\nu_{\text{sub}}(I, \mathcal{W}) = 1/2$.

□

The last simple example shows a fuzzy extension which does not influence the subadditivity of its deterministic pattern.

Example 11.6 Let $(I, \mathcal{V})$ be the 2-players coalition game already processed in the previous two examples. Let $(I, \mathcal{W})$ be its fuzzy extension with κ_I used in Example 11.4, i.e. for all $\boldsymbol{x} \in R^2$

$$\begin{aligned} \kappa_I(\boldsymbol{x}) &= 1 && \text{for} \quad x_1 \cdot x_2 \leq 1, \\ &= 0 && \text{for} \quad (x_1 - 1) \cdot (x_2 - 1) \geq 1, \\ &= 1 - \lambda && \text{for} \quad (x_1 - \lambda) \cdot (x_2 - \lambda) = 1, \ 0 < \lambda < 1. \end{aligned}$$

Let $\mathcal{W}(\{i\})$, $i = 1, 2$, be defined by

$$\begin{aligned} \kappa_{\{i\}}(\boldsymbol{x}) &= 1 && \text{for} \quad x_i \leq 1, \\ &= 1 - x_i && \text{for} \quad 1 < x_i < 2, \\ &= 0 && \text{for} \quad x_i \geq 2. \end{aligned}$$

Then, obviously, $\mathcal{W}^*(\{i\}) = \mathcal{V}^*(\{i\})$, $i = 1, 2$, and $\mathcal{W}^*(I) = \mathcal{V}^*(I)$, i.e.

$$\begin{aligned} \kappa_I^*(\boldsymbol{x}) &= 1 && \text{for} \quad \boldsymbol{x} \in \mathcal{V}^*(I), \\ &= 0 && \text{for} \quad \boldsymbol{x} \notin \mathcal{V}^*(I), \end{aligned}$$

$$\begin{aligned} \kappa_{\{i\}}^*(\boldsymbol{x}) &= 1 && \text{for} \quad \boldsymbol{x} \in \mathcal{V}^*(\{i\}), \\ &= 0 && \text{for} \quad \boldsymbol{x} \notin \mathcal{V}^*(\{i\}). \end{aligned}$$

Consequently, we have

$$\min\left(\kappa_{\{1\}}^*(\boldsymbol{x}), \kappa_{\{2\}}^*(\boldsymbol{x})\right) \leq \kappa_I^*(\boldsymbol{x})$$

for all $\boldsymbol{x} \in R^2$, and $(I, \mathcal{W})$ is imputationally subadditive and fuzzy subadditive with possibility $\nu_{\text{sub}}(I, \mathcal{W}) - 1$.

□

For illustrating the properties of the fuzzy and imputational additivity we employ the results obtained in the previous examples in this section.

Example 11.7 Let $(I, \mathcal{V})$ be a 2-players coalition game which was already considered in Examples 11.1, 11.2 and 11.3, and let $(I, \mathcal{W})$ be its fuzzy extension such that for $\boldsymbol{x} \in R^2$

$$\begin{aligned} \kappa_I(\boldsymbol{x}) &= 1 && \text{for} \quad \boldsymbol{x} \in \mathcal{V}(I), \\ &= 0 && \text{for} \quad \boldsymbol{x} \notin \mathcal{V}(I), \end{aligned}$$

$$\begin{aligned} \kappa_{\{i\}}(\boldsymbol{x}) &= 1 && \text{for} \quad x_i \leq 5, \\ &= 0 && \text{for} \quad x_i \geq 7, \\ &= (7 - x_i)/2 && \text{for} \quad x_i \in (5, 7). \end{aligned}$$

The game $(I, \mathcal{V})$ is superadditive but not subadditive, $(I, \mathcal{W})$ is neither imputationally nor fuzzy superadditive as

$$\min \left(\kappa_{\{1\}}(\boldsymbol{x}), \kappa_{\{2\}}(\boldsymbol{x})\right) > \kappa_I(\boldsymbol{x})$$

for $\boldsymbol{x} \in R^2$ such that $x_1 + x_2 > 5$, $x_1 < 7/2$ and $x_2 < 7/2$. This means that $\nu_{\text{super}}(I, w) = 0$ (analogously to Example 11.3).

On the other hand, we have

$$\begin{aligned} \kappa_I^*(\boldsymbol{x}) &= 1 && \text{for} \quad \boldsymbol{x} \in \mathcal{V}^*(I), \\ &= 0 && \text{for} \quad \boldsymbol{x} \notin \mathcal{V}^*(I), \end{aligned}$$

where

$$\mathcal{V}^*(I) = \{\boldsymbol{x} \in R^2 : x_1 + x_2 \geq 5\} \cup \{\boldsymbol{x} \in R^2 : x_1 > 5\} \cup \{\boldsymbol{x} \in R^2 : x_2 > 5\},$$

and for $i = 1, 2$

$$\begin{aligned} \kappa_{\{i\}}^*(\boldsymbol{x}) &= 1 && \text{for} \quad x_i \geq 5/2, \\ &= 0 && \text{for} \quad x_i < 5/2. \end{aligned}$$

This means that

$$\kappa_i^*(\boldsymbol{x}) \geq \min \left(\kappa_{\{1\}}^*(\boldsymbol{x}), \kappa_{\{2\}}^*(\boldsymbol{x})\right),$$

$(I, \mathcal{W})$ is imputationally subadditive and $\nu_{\text{sub}}(I, \mathcal{W}) = 1$.

Consequently, $(I, \mathcal{W})$ is neither imputationally additive nor fuzzy additive with any positive possibility,

$$\nu_{\text{addit}}(I, \mathcal{W}) = \min \left(\nu_{\text{super}}(I, \mathcal{W}), \nu_{\text{sub}}(I, \mathcal{W})\right) = 0,$$

and $\nu_{\text{sub}}(I, \mathcal{W}) > \nu_{\text{super}}(I, \mathcal{W})$.

□

Example 11.8 Let us consider now the 2-players coalition game $(I, \mathcal{V})$ and its fuzzy extension already dealt with in Example 11.5.

Using its results we know that $(I, \mathcal{W})$ is not imputationally subadditive and

$$\nu_{\text{sub}}(I, \mathcal{W}) = 1/2.$$

Regarding the superadditivity of $(I, \mathcal{W})$, it is easy to see that for all $x \in R^2$

$$\min\left(\kappa_{\{1\}}(x), \kappa_{\{2\}}(x)\right) \leq \kappa_I(x)$$

which means that $(I, \mathcal{W})$ is imputationally superadditive and $\nu_{\text{super}}(I, \mathcal{W}) = 1$.

Let us note that the deterministic game $(I, \mathcal{V})$ is both superadditive and subadditive, and that it is in consequence also additive.

Its fuzzy extension $(I, \mathcal{W})$ is not imputationally additive but it is fuzzy additive with possibility

$$\nu_{\text{addit}}(I, \mathcal{W}) = 1/2,$$

and $\nu_{\text{sub}}(I, \mathcal{W}) < \nu_{\text{super}}(I, \mathcal{W})$.

□

Example 11.9 Here, we deal with the coalition game $(I, \mathcal{V})$ and its fuzzy extension $(I, \mathcal{W})$ considered in Example 11.6. It means that $\nu_{\text{sub}}(I, \mathcal{W}) = 1$. On the other hand, it is not difficult to see that for all $x \in R^2$

$$\kappa_I(x) \geq \min\left(\kappa_{\{1\}}(x), \kappa_{\{2\}}(x)\right)$$

and, consequently, $\nu_{\text{super}}(I, \mathcal{W}) = 1$, and $(I, \mathcal{W})$ is also imputationally superadditive.

This means that

$$\nu_{\text{super}}(I, \mathcal{W}) = \nu_{\text{sub}}(I, \mathcal{W}) = \nu_{\text{addit}}(I, \mathcal{W}) = 1.$$

□

The statements on additivity-like properties, i.e. superadditivity, subadditivity and additivity, of fuzzy coalition games presented in this chapter show that both approaches to their definition – the imputational and the fuzziness representing ones – essentially differ from their counterparts in the games with side-payments.

A mutual comparison of both types of coalition games, with and without side-payments will be possible for so-called linear coalition games without side-payments which are briefly discussed in Chapter 13.

12. Fuzzy Core and Effectivity in Games Without Side–Payments

The core $\boldsymbol{C}$ of the deterministic coalition game without side-payments $(I, \mathcal{W})$ is, due to (3.21), a set of imputations which are accessible for some coalition structure (in the superadditive case for the coalition of all players), and which cannot be protested by any coalition.

In the formalism used in Chapter 3, if an imputation $\boldsymbol{x}$ is to belong to the core $\boldsymbol{C}$, then there has to exist a coalition structure $(K_1, \ldots, K_m)$ such that $\boldsymbol{x} \in \mathcal{V}(K_j)$ for all $j = 1, \ldots, m$, and $\boldsymbol{x}$ has to belong to $\mathcal{V}^*(K)$ for all coalitions $K \subset I$. This principle will also be preserved for fuzzy coalition games without side-payments.

This means that the *fuzzy core* $\boldsymbol{C}_F$ *of* $(I, \mathcal{W})$ extending a coalition game without side-payments $(I, \mathcal{V})$ is a fuzzy subset of R^I with the membership function $\gamma_{\boldsymbol{C}} : R^I \to [0,1]$. This membership function can be derived from the membership functions of $\mathcal{W}(K)$ and $\mathcal{W}^*(K)$ in the following way.

The possibility that $\boldsymbol{x}$ is accessible by a coalition structure $\mathcal{K}$, i.e. $\boldsymbol{x}$ belongs to all $\mathcal{W}(K)$ for $K \in \mathcal{K}$, is

$$\kappa_{\mathcal{K}}(\boldsymbol{x}) = \min(\kappa_K(\boldsymbol{x}) : K \in \mathcal{K}).$$

Let us note that $\kappa_{\mathcal{K}}(\boldsymbol{x})$ is the membership function of a fuzzy subset of R^I, that there always exists $\boldsymbol{x} \in R^I$ such that $\kappa_{\mathcal{K}}(\boldsymbol{x}) = 1$, and that $\kappa_{\mathcal{K}}$ generally fulfills also other properties of the membership functions of the sets $\mathcal{W}(K)$.

Namely, if we denote

$$\mathcal{V}(\mathcal{K}) = \bigcap\nolimits_{K \in \mathcal{K}} \mathcal{V}(K)$$

then $\kappa_{\mathcal{K}}(\boldsymbol{x}) = 1$ for all $\boldsymbol{x} \in \mathcal{V}(\mathcal{K})$; if $\boldsymbol{x}, \boldsymbol{y} \in R^I$, $x_i \geq y_i$ for all $i \in I$, then $\kappa_{\mathcal{K}}(\boldsymbol{x}) \leq \kappa_{\mathcal{K}}(\boldsymbol{y})$; there exists $\boldsymbol{x} \in \mathbb{R}^I$ such that $\kappa_{\mathcal{K}}(\boldsymbol{x}) = 0$; and also the set $\{\boldsymbol{x} \in R^I : \kappa_{\mathcal{K}}(\boldsymbol{x}) = 1\}$ is closed.

The possibility that there exists a coalition structure for which $\boldsymbol{x}$ is accessible will be denoted

$$\kappa_{\boldsymbol{C}}(\boldsymbol{x}) = \max(\kappa_{\mathcal{K}}(\boldsymbol{x}) : \ \mathcal{K} \text{ is a coalition structure}). \tag{12.1}$$

On the other hand, the possibility that $\boldsymbol{x}$ cannot be protested by any coalition, i.e. it cannot be significantly dominated by any accessible imputation, is

$$\lambda_{\boldsymbol{C}} = \min(\kappa_K^*(\boldsymbol{x}) : K \subset I). \tag{12.2}$$

Then it is easy to define the membership function $\gamma_{\boldsymbol{C}}$ by

$$\gamma_{\boldsymbol{C}}(\boldsymbol{x}) = \min(\kappa_{\boldsymbol{C}}(\boldsymbol{x}), \lambda_{\boldsymbol{C}}(\boldsymbol{x})), \quad \boldsymbol{x} \in R^I. \tag{12.3}$$

Evidently, if the fuzzy coalition game $(I, \mathcal{W})$ is imputationally superadditive in the sense of $\mathcal{W}(K \cup L) \supset \mathcal{W}(K) \cap \mathcal{W}(L)$ for any disjoint pair of coalitions K, L, then the coalition structure in which the imputations forming the core $\boldsymbol{C}_F$ are realized is the one all-players coalition I.

Then (12.3) turns into

$$\gamma_{\boldsymbol{C}}(\boldsymbol{x}) = \min(\kappa_I(\boldsymbol{x}), \lambda_{\boldsymbol{C}}(\boldsymbol{x})) \tag{12.4}$$

which is equivalent to

$$\boldsymbol{C}_F = \mathcal{W}(I) \cap \left(\bigcap\nolimits_{L \subset I} \mathcal{W}^*(L)\right).$$

It is not difficult to verify (cf. [24, 35]) that this fuzzy core $\boldsymbol{C}_F$ of $(I, \mathcal{W})$ is a fuzzy extension of the core $\boldsymbol{C}$ of its deterministic pattern $(I, \mathcal{V})$. Namely, if $(I, \mathcal{W})$ is a fuzzy extension of $(I, \mathcal{V})$ and $\boldsymbol{C}$ is the core of $(I, \mathcal{V})$ then, in the general case, $\gamma_C(\boldsymbol{x}) > 0$ for $\boldsymbol{x} \in \boldsymbol{C}$ and $\gamma_{\boldsymbol{C}}(\boldsymbol{x}) < 1$ if $\boldsymbol{x} \notin \boldsymbol{C}$.

In a special case when for any coalition $K \subset I$, $\kappa_K(\boldsymbol{x})$ is strictly decreasing on the set

$$\{\boldsymbol{x} \in R^I : \kappa_K(\boldsymbol{x}) \in (0,1)\}$$

in the sense that $\kappa_K(\boldsymbol{x}) < \kappa_K(\boldsymbol{y})$ if $\boldsymbol{x} \operatorname{dom}_K \boldsymbol{y}$, then $\gamma_{\boldsymbol{C}}(\boldsymbol{x}) = 1$ for $\boldsymbol{x} \in C$.

Moreover, if for all $K \subset I$ the membership function $\mu_K(\boldsymbol{x}) = 1$ iff $\boldsymbol{x} \in \mathcal{V}(K)$, and the previous condition of strict monotonicity is fulfilled, then $\gamma_{\boldsymbol{C}}(\boldsymbol{x}) = 1$ iff $\boldsymbol{x} \in \boldsymbol{C}$.

As it was already mentioned in the previous chapters, under the above assumptions the strict monotonicity of κ_K implies the degeneration of the fuzzy superoptimum set $\mathcal{W}^*$ into its deterministic pattern $\mathcal{V}^*$, namely, $\kappa_K^*(\boldsymbol{x}) = 1$ for $\boldsymbol{x} \in \mathcal{V}^*(K)$ and $\kappa_K^*(\boldsymbol{x}) = 0$ for $\boldsymbol{x} \notin \mathcal{V}^*(K)$.

Then the fuzzy core $\boldsymbol{C}_F$ is, in fact, determined by the fuzzy sets $\mathcal{W}(K)$ where $K \in \mathcal{K}$ and $\mathcal{K}$ are the coalition structures in (12.1), and by the intersection

$$\bigcap\nolimits_{L \subset I} \mathcal{V}^*(L)$$

and, for $\boldsymbol{x}$ outside the certainly accessible imputations, i.e.

$$\boldsymbol{x} \notin \bigcup\nolimits_{\text{all } \mathcal{K}} \left(\bigcap\nolimits_{K \in \mathcal{K}} \mathcal{V}(K)\right), \quad \mathcal{K} \text{ are coalition structures,}$$

the possibility value $\gamma_{\boldsymbol{C}}(\boldsymbol{x})$ is fully determined by its accessibility for coalition structures, i.e. by the values $\kappa_{\boldsymbol{C}(\boldsymbol{x})}$.

The possibility that the fuzzy coalition game $(I, \mathcal{W})$ has a non-empty core can be denoted by

$$\gamma(I, \mathcal{W}) = \sup\left(\gamma_{\boldsymbol{C}}(\boldsymbol{x}) : \boldsymbol{x} \in R^I\right)$$

and the results mentioned above mean that

$$\gamma(I, \mathcal{W}) > 0 \text{ for } \boldsymbol{C} \neq \emptyset \quad \text{and} \quad \gamma(I, \mathcal{W}) < 1 \text{ for } \boldsymbol{C} = \emptyset.$$

Also the concept of effectivity of coalitions, whose deterministic version was presented in Chapter 3, Section 3.2, and whose analogue for games with side-payments was already presented in Chapter 6, exhibits quite natural and expectable properties.

We say that a coalition $K \subset I$ is *fuzzy effective* in the game $(I, \mathcal{W})$ if there exists an imputation $\boldsymbol{x} \in R^I$ which is accessible for K, i.e. $\boldsymbol{x} \in \mathcal{W}(K)$, and which cannot be protested by any of its subcoalitions, i.e. $\boldsymbol{x} \in \mathcal{W}^*(J)$ for all $J \subset I$.

The possibility that K is effective is then given by the possibilities of those phenomena for some imputations. Namely, the possibility $\eta(K)$ that K is effective in $(I, \mathcal{W})$ is

$$\eta(K) = \sup_{\boldsymbol{x} \in R^I} \left[\min_{J \subset K}(\kappa_K(\boldsymbol{x}), \kappa^*_J(\boldsymbol{x}))\right]. \tag{12.5}$$

Of course, every 1-player coalition $\{i\}$, $i \in I$, is effective with possibility $\eta(\{i\}) = 1$. The results presented in Chapter 11 and (12.4) imply that if $(I, \mathcal{W})$ is superadditive in the sense that $\mathcal{W}(K \cup L) \supset \mathcal{W}(K) \cap \mathcal{W}(L)$ for any pair of disjoint coalitions K, L, then the coalition I is effective with possibility

$$\eta(I) = \sup(\gamma_{\boldsymbol{C}}(\boldsymbol{x}) : \boldsymbol{x} \in R^I).$$

It also means that if $(I, \mathcal{W})$ is a fuzzy extension of $(I, \mathcal{V})$, the core $\boldsymbol{C}$ of $(I, \mathcal{V})$ is non-empty and if $(I, \mathcal{V})$ is superadditive, then $\eta(I) > 0$. In a special case of 2-players coalition $\{i, j\} \subset I$, $i \neq j$, fulfilling $\mathcal{V}(\{i, j\}) \supset \mathcal{V}(\{i\}) \cap \mathcal{V}(\{j\})$, the effectivity is positive, i.e. $\eta(\{i, j\}) > 0$. If, moreover, $\kappa_{\{i\}}(\boldsymbol{x})$, and $\kappa_{\{j\}}(\boldsymbol{x})$ are strictly monotone in the sense that for

$$\boldsymbol{x}, \boldsymbol{y} \in \{\boldsymbol{z} \in R^I : \kappa_{\{i\}}(\boldsymbol{z}) \in (0, 1)\}$$

the inequality $x_i > y_i$ implies $\kappa_{\{i\}}(\boldsymbol{x}) < \kappa_{\{i\}}(\boldsymbol{y})$ and analogously for

$$\boldsymbol{x}, \boldsymbol{y} \in \{\boldsymbol{z} \in R^I : \kappa_{\{j\}}(\boldsymbol{z}) \in (0, 1)\}$$

the inequality $x_j > y_j$ implies $\kappa_{\{j\}}(\boldsymbol{x}) < \kappa_{\{j\}}(\boldsymbol{y})$, then it can be easily verified that $\eta(\{i, j\}) = 1$.

Having introduced the concept of effectivity of coalitions, it is rational to consider the possibility of forming particular coalition structures. They are, generally speaking, as possible as the coalitions which form them. Namely, if $\mathcal{K}$ is a coalition structure, then we can introduce the concept of its *fuzzy effectivity* and say that $\mathcal{K}$ is fuzzy effective with possibility $\eta(\mathcal{K})$, where

$$\eta(\mathcal{K}) = \min(\eta(K) : K \in \mathcal{K}). \tag{12.6}$$

It is useful to note that the high value of possibility $\eta(\mathcal{K})$ does not mean that $\mathcal{K}$ can be expected as a highly possible result of bargaining. In the previous paragraphs, for example, we have seen relatively simple cases of games where $\eta(I) = 1$ (and, consequently, $\eta(\{I\}) = 1$, where $\{I\}$ is the coalition structure of single all-players coalition) and we already know that also that coalition structure consisting exclusively of all 1-player coalitions $\{i\}$, $i \in I$, is also effective with possibility $\eta((\{i\})_{i\in I}) = 1$.

In many superadditive games the coalition I is the single expectable result of bargaining while the coalition structure $(\{i\})_{i\in I}$ often rather appears as a starting point of the negotiation process.

The effectivity of a coalition structure only means that it can appear during the process of bargaining where it may (but need not) be substituted by another one, either by means of the union of some coalitions or by a re-grouping of players as a result of negotiations. Only the coalition of all players I, when being certainly effective, i.e. $\eta(I) = 1$, represents a definite results of bargaining.

Let $(I, \mathcal{W})$ be a fuzzy coalition game which is imputationally superadditive in the sense that $\mathcal{W}(K \cup L) \supset \mathcal{W}(K) \cap \mathcal{W}(L)$ for any disjoint $K, L \subset I$.

Then for any coalition structure $\mathcal{K}$, we obviously have

$$\eta(\{I\}) \geq \eta(\mathcal{K}).$$

If we want to illustrate the above concepts by a transparent example, it is most transparent to do so by a 2-players game where the eventual graphical imagination can help understand.

Example 12.1 Let us consider a 2-players coalition game $(I, \mathcal{V})$ with $I = \{1, 2\}$,

$$\mathcal{V}(\{1\}) = \{\boldsymbol{x} \in R^2 : x_1 \leq 1\}, \qquad \mathcal{V}(\{2\}) = \{\boldsymbol{x} \in R^2 : x_2 \leq 2\},$$
$$\mathcal{V}(I) = \{\boldsymbol{x} \in R^2 : x_1 + x_2 \leq 5\}.$$

Let us note that this game has a non-empty core

$$\boldsymbol{C} = \left\{\boldsymbol{x} \in R^2 : x_1 + x_2 = 5, x_1 \geq 1, x_2 \geq 2\right\}.$$

We can consider its fuzzy extension $(I, \mathcal{W})$ where

$$\begin{aligned}\kappa_{\{1\}}(\boldsymbol{x}) &= 1 && \text{for} \quad x_1 \le 1,\\ &= 2 - x_1 && \text{for} \quad 1 < x_1 < 2,\\ &= 0 && \text{for} \quad x_1 \ge 0,\end{aligned}$$

$$\begin{aligned}\kappa_{\{2\}}(\boldsymbol{x}) &= 1 && \text{for} \quad x_2 \le 2,\\ &= 3 - x_2 && \text{for} \quad 2 < x_2 < 3,\\ &= 0 && \text{for} \quad x_2 \ge 3,\end{aligned}$$

$$\begin{aligned}\kappa_I(\boldsymbol{x}) &= 1 && \text{for} \quad x_1 + x_2 \le 5,\\ &= 6 - (x_1 + x_2) && \text{for} \quad 5 < x_1 + x_2 < 6,\\ &= 0 && \text{for} \quad x_1 + x_2 \ge 6.\end{aligned}$$

With the experience gained in Example 10.1 and Example 10.7 we can easily find that

$$\mathcal{W}^*(\{1\}) = \mathcal{V}^*(\{1\}), \quad \mathcal{W}^*(\{2\}) = \mathcal{V}^*(\{2\}), \quad \mathcal{W}^*(I) = \mathcal{V}^*(I)$$

in the sense that

$$\begin{aligned}\kappa^*_{\{1\}}(\boldsymbol{x}) &= 1 \quad \text{for} \quad x_1 \ge 1, & \kappa^*_{\{2\}}(\boldsymbol{x}) &= 1 \quad \text{for} \quad x_2 \ge 2,\\ &= 0 \quad \text{for} \quad x_1 < 1, & &= 0 \quad \text{for} \quad x_2 < 2,\end{aligned}$$

$$\begin{aligned}\kappa_I(\boldsymbol{x}) &= 1 \quad \text{for} \quad x_1 + x_2 \ge 5,\\ &= 0 \quad \text{for} \quad x_1 + x_2 < 5.\end{aligned}$$

and the fuzzy Pareto optimum $\mathcal{Q}(I)$ is the fuzzy set with membership function

$$\begin{aligned}\rho_I(\boldsymbol{x}) &= 0 && \text{for} \quad x_1 + x_2 < 5 \text{ or } x_1 + x_2 > 6,\\ &= 6 - (x_1 + x_2) && \text{for} \quad 5 \le x_1 + x_2 \le 6.\end{aligned}$$

The fuzzy coalition game $(I, \mathcal{W})$ is imputationally superadditive and it is easy to see that its fuzzy core $\boldsymbol{C}_F = \mathcal{Q}(I) \cap \mathcal{W}^*(\{1\}) \cap \mathcal{W}^*(\{2\})$, i.e.

$$\begin{aligned}\gamma_{\boldsymbol{C}}(\boldsymbol{x}) &= 0 && \text{for} \quad x_1 + x_2 < 5 \text{ or } x_1 + x_2 > 6\\ & && \qquad \text{or } x_1 < 1 \text{ or } x_2 < 2\\ &= 6 - (x_1 + x_2) && \text{for} \quad \boldsymbol{x} \in \{\boldsymbol{y} \in R^2 : 5 \le y_1 + y_2 \le 6\} \cap\\ & && \qquad \cap \{\boldsymbol{y} \in R^2 : y_1 \ge 1\} \cap\\ & && \qquad \cap \{\boldsymbol{y} \in R^2 : y_1 \ge 2\},\end{aligned}$$

as for the coalition structure $\mathcal{K} = (I)$, $\kappa_{\mathcal{K}}(\boldsymbol{x}) = \kappa_I(\boldsymbol{x})$, and for the coalition structure $\mathcal{L} = (\{1\}, \{2\})$

$$\begin{aligned}\kappa_{\mathcal{L}}(\boldsymbol{x}) &= 1 && \text{for} \quad \boldsymbol{x} \in \{\boldsymbol{y} \in R^2 : y_1 \le 1\} \cap\\ & && \qquad \cap \{\boldsymbol{y} \in R^2 : y_2 \le 2\},\\ &= 2 - x_1 && \text{for} \quad \boldsymbol{x} \in \{\boldsymbol{y} \in R^2 : 1 < y_1 < 2\} \cap\\ & && \qquad \cap \{\boldsymbol{y} \in R2 : y_2 \le y_1 + 1\}\\ &= 2 - x_2 && \text{for} \quad \boldsymbol{x} \in \{\boldsymbol{y} \in R^2 : 2 < y_2 < 3\} \cap\\ & && \qquad \cap \{\boldsymbol{y} \in R2 : y_1 \le y_2 - 1\}\\ &= 0 && \text{otherwise,}\end{aligned}$$

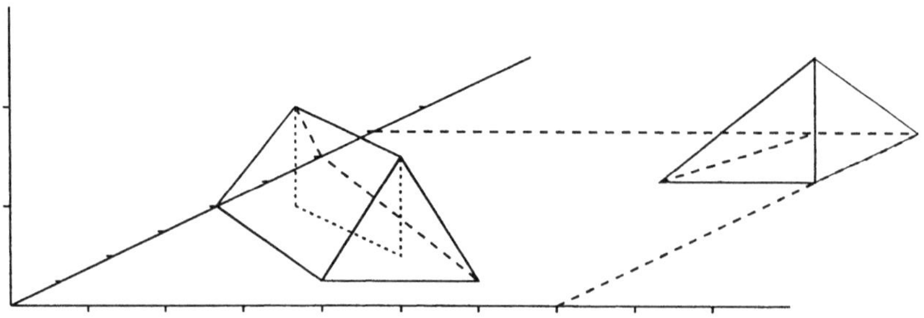

Fig. 12.1. Fuzzy cores (membership functions γ_C) of games in Example 12.1 (C1) and Example 12.2 (C2)

and, consequently,

$$\kappa_C(\boldsymbol{x}) = \kappa_I(\boldsymbol{x}).$$

Moreover, it is easy to see that

$$\begin{aligned} \lambda_C(\boldsymbol{x}) &= 1 \quad \text{for} \quad && \boldsymbol{x} \in \{\boldsymbol{y} \in R^2 : y_1 + y_2 \geq 5\} \cap \\ & && \cap \{\boldsymbol{y} \in R^2 : y_1 \geq 1\} \cap \\ & && \cap \{\boldsymbol{y} \in R^2 : y_2 \geq 2\}, \\ &= 0 \quad \text{otherwise.} && \end{aligned}$$

The fuzzy core $\boldsymbol{C}_F$ is shown in Figure 12.1. The previous calculation also shows that

$$\gamma(I, \mathcal{W}) = 1,$$

that the coalition I is effective with possibility

$$\eta(I) = 1$$

(let us remember that always $\eta(\{1\}) = \eta(\{2\}) = 1$), and that for both coalition structures $\mathcal{K}, \mathcal{L}$ introduced above

$$\eta(\mathcal{K}) = \eta(\mathcal{L}) = 1.$$

□

Example 12.2 Let $(I, \mathcal{V})$ be a coalition game with $I = \{1, 2\}$

$$\mathcal{V}(\{1\}) = \{\boldsymbol{x} \in R^2 : x_1 \leq 6\}, \qquad \mathcal{V}(\{2\}) = \{\boldsymbol{x} \in R^2 : x_2 \leq 6\},$$
$$\mathcal{V}(I) = \{\boldsymbol{x} \in R^2 : x_1 + x_2 \leq 9\}.$$

and let $(I, \mathcal{W})$ be its fuzzy extension such that for $i = 1, 2$

$$\begin{aligned}\kappa_{\{i\}}(\boldsymbol{x}) &= 1 && \text{for} \quad x_i \leq 6,\ i=1,2\\ &= 7-x_i && \text{for} \quad 6 < x_i < 7,\\ &= 0 && \text{for} \quad x_i \geq 7,\end{aligned}$$

$$\begin{aligned}\kappa_I(\boldsymbol{x}) &= 1 && \text{for} \quad x_1+x_2 \leq 9,\\ &= 10-(x_1+x_2) && \text{for} \quad 9 < x_1+x_2 < 10,\\ &= 0 && \text{for} \quad x_1+x_2 \geq 10.\end{aligned}$$

Then, it is easy to verify that

$$\begin{aligned}\kappa^*_{\{i\}}(\boldsymbol{x}) &= 0 \quad \text{for} \quad x_i < 6, & \kappa^*_I(\boldsymbol{x}) &= 0 \quad \text{for} \quad x_i < 0,\\ &= 1 \quad \text{for} \quad x_i \geq 6, & &= 1 \quad \text{for} \quad x_i \geq 9,\end{aligned}$$

The Pareto optima $\mathcal{Q}(\{i\})$ and $\mathcal{Q}(I)$ fulfill

$$\begin{aligned}\rho_I(\boldsymbol{x}) &= 0 && \text{for} \quad x_1+x_2 < 9 \text{ or } x_1+x_1 \geq 10,\\ &= 10-(x_1+x_2) && \text{for} \quad 9 \leq x_1+x_2 < 10,\end{aligned}$$

$$\begin{aligned}\rho_{\{i\}}(\boldsymbol{x}) &= 0 && \text{for} \quad x_i < 6 \text{ or } x_i \geq 7,\\ &= 7-x_i && \text{for} \quad 6 \leq x_i < 7.\end{aligned}$$

For the coalition structures $\mathcal{K} = (I)$, $\mathcal{L} = (\{1\},\{2\})$, $\kappa_{\mathcal{K}}(\boldsymbol{x}) = \kappa_I(\boldsymbol{x})$, and

$$\begin{aligned}\kappa_{\mathcal{L}}(\boldsymbol{x}) &= 1 && \text{for} && \boldsymbol{x} \in \{\boldsymbol{y} \in R^2 : y_1 \leq 6\} \cap \\ & && && \cap \{\boldsymbol{y} \in R^2 : y_2 \leq 6\},\\ &= 7-x_1 && \text{for} && \boldsymbol{x} \in \{\boldsymbol{y} \in R^2 : 7 < y_1 < 6\} \cap \\ & && && \cap \{\boldsymbol{y} \in R^2 : y_2 \leq y_1\},\\ &= 7-x_2 && \text{for} && \boldsymbol{x} \in \{\boldsymbol{y} \in R^2 : 6 < y_1 < 7\} \cap \\ & && && \cap \{\boldsymbol{y} \in R^2 : y_1 \leq y_2\},\\ &= 0 && \text{otherwise}, && \end{aligned}$$

$$\begin{aligned}\kappa_{\boldsymbol{C}}(\boldsymbol{x}) &= \max\left(\kappa_{\mathcal{K}}(\boldsymbol{x}), \kappa_{\mathcal{L}}(\boldsymbol{x})\right)\\ \lambda_{\boldsymbol{C}}(\boldsymbol{x}) &= \min\left(\kappa^*_I(\boldsymbol{x}), \kappa^*_{\{1\}}(\boldsymbol{x}), \kappa^*_{\{2\}}(\boldsymbol{x})\right) = \min\left(\kappa^*_{\{1\}}(\boldsymbol{x}), \kappa^*_{\{2\}}(\boldsymbol{x})\right)\end{aligned}$$

i.e.

$$\begin{aligned}\lambda_{\boldsymbol{C}}(\boldsymbol{x}) &= 1 \quad \text{for} \quad \boldsymbol{x} \in \{\boldsymbol{y} \in R^2 : y_1 \geq 7\} \cap \{\boldsymbol{y} \in R^2 : y_2 \geq 6\}\\ &= 0 \quad \text{otherwise.}\end{aligned}$$

Then, we have

$$\gamma_{\boldsymbol{C}}(\boldsymbol{x}) = \min\left(\rho_{\{1\}}(\boldsymbol{x}), \rho_{\{2\}}(\boldsymbol{x})\right),$$

i.e.

$$\begin{aligned}\gamma_{\boldsymbol{C}}(\boldsymbol{x}) &= 0 && \text{for} \quad \boldsymbol{x} \in \{\boldsymbol{y} \in R^2 : y_1 < 6\} \cup \{\boldsymbol{y} \in R^2 : y_2 < 6\} \\ &&& \qquad \cup \{\boldsymbol{y} \in R^2 : y_1 > 7\} \cup \{\boldsymbol{y} \in R^2 : y_2 > 7\}, \\ &= 7 - x_1 && \text{for} \quad \boldsymbol{x} \in \{\boldsymbol{y} \in R^2 : 6 \le y_1 \le 7\} \cap \{\boldsymbol{y} \in R^2 : 6 \le y_2 \le y_1\}, \\ &= 7 - x_2 && \text{for} \quad \boldsymbol{x} \in \{\boldsymbol{y} \in R^2 : 6 \le y_2 \le 7\} \cap \{\boldsymbol{y} \in R^2 : 6 \le y_1 \le y_2\}.\end{aligned}$$

Obviously,

$$\gamma(I, \mathcal{W}) = 1 \quad \text{as} \quad \gamma_{\boldsymbol{C}}(\boldsymbol{x}) = 1 \quad \text{for } \boldsymbol{x} = (6, 6).$$

The fuzzy core of $(I, \mathcal{W})$ is also shown in Figure 12.1.

It is not difficult to see, that the coalition I in this game is not fuzzy effective at all, $\eta(I) = 0$, which means that $\eta(\mathcal{K}) = 0$ and there is the zero possibility that this coalition can appear during the bargaining process.

□

Example 12.3 Let us consider a 2-players coalition game $(I, \mathcal{V})$ such that $I = \{1, 2\}$,

$$\begin{aligned}\mathcal{V}(\{i\}) &= \{\boldsymbol{x} \in R^2 : x_i \le 1\}, \quad i = 1, 2, \\ \mathcal{V}(I) &= \{\boldsymbol{x} \in R^2 : x_1 + x_2 \le 5\}) \cap \{\boldsymbol{x} \in R^2 : x_1 \le 5\} \cap \\ & \quad \cap \{\boldsymbol{x} \in R^2 : x_2 \le 5\}.\end{aligned}$$

It is evident that all coalitions in this deterministic game are effective and its core $\boldsymbol{C}$ is realized by the all-players coalition,

$$\boldsymbol{C} = \{\boldsymbol{x} \in R^2 : x_1 + x_2 = 5, \ x_1 \ge 1, \ x_2 \ge 1\}.$$

Let us consider the fuzzy extension $(I, \mathcal{W})$ of $(I, \mathcal{V})$ such that for $\boldsymbol{x} \in R^2$

$$\begin{aligned}\kappa_{\{i\}}(\boldsymbol{x}) &= 1 && \text{for} \quad x_i \le 1, \ i = 1, 2, \\ &= 1/2 && \text{for} \quad 1 < x_i \le 3, \\ &= 0 && \text{for} \quad x_i > 3,\end{aligned}$$

$$\begin{aligned}\kappa_I(\boldsymbol{x}) &= 1 && \text{for} \quad \boldsymbol{x} \in \{\boldsymbol{y} \in R^2 : y_1 + y_2 \le 5\} \cap \{\boldsymbol{y} \in R^2 : y_1 \le 5\} \cap \\ &&& \qquad \cap \{\boldsymbol{y} \in R^2 : y_2 \le 5\} = \mathcal{V}(I), \\ &= 1/2 && \text{for} \quad \boldsymbol{x} \in \{\boldsymbol{y} \in R^2 : 5 < y_1 + y_2 \le 6\} \cap \\ &&& \qquad \cap \{\boldsymbol{y} \in R^2 : y_1 \le 5\} \cap \\ &&& \qquad \cap \{\boldsymbol{y} \in R^2 : y_2 \le 5\}, \\ &= 0 && \text{for} \quad \boldsymbol{x} \in \{\boldsymbol{y} \in R^2 : y_1 + y_2 > 6\} \cup \\ &&& \qquad \cup \{\boldsymbol{y} \in R^2 : y_1 > 5\} \cup \{\boldsymbol{y} \in R^2 : y_2 > 5\}.\end{aligned}$$

Analogously to Example 10.3 and Example 10.8, we can see that, for $\boldsymbol{x} \in R^2$, $i = 1, 2$, there is

$$
\begin{aligned}
\kappa_I^*(\boldsymbol{x}) &= 0 && \text{for} \quad \boldsymbol{x} \in \{\boldsymbol{y} \in R^2 : y_1 + y_2 < 5\} \cap \{\boldsymbol{y} \in R^2 : y_1 \leq 5\} \cap \\
&&& \qquad\qquad \cap \{\boldsymbol{y} \in R^2 : y_2 \leq 5\}, \\
&= 1 && \text{for} \quad \boldsymbol{x} \in \{\boldsymbol{y} \in R^2 : y_1 + y_2 = 5\} \cap \{\boldsymbol{y} \in R^2 : y_1 \leq 5\} \cap \\
&&& \qquad\qquad \cap \{\boldsymbol{y} \in R^2 : y_2 \leq 5\} \\
&= 1/2 && \text{for} \quad \boldsymbol{x} \in \{\boldsymbol{y} \in R^2 : 5 < y_1 + y_2 < 6\} \cap \\
&&& \qquad\qquad \cap \{\boldsymbol{y} \in R^2 : y_1 \leq 5\} \cap \{\boldsymbol{y} \in R^2 : y_2 \leq 5\}, \\
&= 1 && \text{for} \quad \boldsymbol{x} \in \{\boldsymbol{y} \in R^2 : y_1 + y_2 = 6\} \cap \{\boldsymbol{y} \in R^2 : y_1 \leq 5\} \cap \\
&&& \qquad\qquad \cap \{\boldsymbol{y} \in R^2 : y_2 \leq 5\}, \\
&= 1 && \text{for} \quad \boldsymbol{x} \in \{\boldsymbol{y} \in R^2 : y_1 + y_2 > 6\} \cup \{\boldsymbol{y} \in R^2 : y_1 > 5\} \cup \\
&&& \qquad\qquad \cup \{\boldsymbol{y} \in R^2 : y_2 > 5\},
\end{aligned}
$$

$$
\begin{aligned}
\kappa_{\{i\}}^*(\boldsymbol{x}) &= 0 && \text{for} \quad x_1 < 1, \\
&= 1 && \text{for} \quad x_i = 1, \\
&= 1/2 && \text{for} \quad 1 < x_i < 3, \\
&= 1 && \text{for} \quad x_i \geq 3.
\end{aligned}
$$

Then, it is possible to compute $\lambda_{\mathcal{C}}(\boldsymbol{x}) = \min(\kappa_{\{1\}}^*(\boldsymbol{x}), \kappa_{\{2\}}^*(\boldsymbol{x}), \kappa_I^*(\boldsymbol{x}))$. Its structure is rather complicated, namely

$$
\begin{aligned}
\lambda_{\mathcal{C}}(\boldsymbol{x}) &= 0 && \text{for} \quad \boldsymbol{x} \in \{\boldsymbol{y} \in R^2 : y_1 + y_2 < 5\} \cup \{\boldsymbol{y} \in R^2 : y_1 < 1\} \cup \\
&&& \qquad\qquad \cup \{\boldsymbol{y} \in R^2 : y_2 < 1\}, \\
&= 1 && \text{for} \quad \boldsymbol{x} \in \{\boldsymbol{y} \in R^2 : y_1 + y_2 = 5\} \cap \{\boldsymbol{y} \in R^2 : y_1 \geq 1\} \cap \\
&&& \qquad\qquad \cap \{\boldsymbol{y} \in R^2 : y_2 \geq 1\}, \\
&= 1/2 && \text{for} \quad \boldsymbol{x} \in \{\boldsymbol{y} \in R^2 : 5 < y_1 + y_2 < 6\} \cap \\
&&& \qquad\qquad \cap \{\boldsymbol{y} \in R^2 : y_1 \geq 1\} \cap \\
&&& \qquad\qquad \cap \{\boldsymbol{y} \in R^2 : y_2 \geq 1\}, \\
&= 1/2 && \text{for} \quad \boldsymbol{x} \in \{\boldsymbol{y} \in R^2 : y_1 + y_2 = 6\} \cap \\
&&& \qquad\qquad \cap \{\boldsymbol{y} \in R^2 : 1 < y_1 < 3\} \cap \\
&&& \qquad\qquad \cap \{\boldsymbol{y} \in R^2 : 1 < y_2 < 3\}, \\
&= 1 && \text{for} \quad \boldsymbol{x} \in \{(1,5), (5,1), (3,3)\} \\
&= 1/2 && \text{for} \quad \boldsymbol{x} \in (\{\boldsymbol{y} \in R^2 : y_1 + y_2 > 6\} \cap \\
&&& \qquad\qquad \cap \{\boldsymbol{y} \in R^2 : 1 \leq y_1 < 3\}) \cup \\
&&& \qquad\qquad \cup (\{\boldsymbol{y} \in R^2 : y_1 + y_2 > 6\} \cap \\
&&& \qquad\qquad \cap \{\boldsymbol{y} \in R^2 : 1 \leq y_2 < 3\}), \\
&= 1 && \text{for} \quad \boldsymbol{x} \in \{\boldsymbol{y} \in R^2 : y_1 \geq 3\} \cap \{\boldsymbol{y} \in R^2 : y_2 \geq 3\}.
\end{aligned}
$$

On the other hand, if we denote the coalition structures $\mathcal{K} = (I)$, $\mathcal{L} = (\{1\}, \{2\})$, then – for all $\boldsymbol{x} \in R^2$ – we have

$$
\begin{aligned}
\kappa_{\mathcal{K}}(\boldsymbol{x}) &= \kappa_I(\boldsymbol{x}), \\
\kappa_{\mathcal{L}}(\boldsymbol{x}) &= 1 \qquad \text{for} \quad \boldsymbol{x} \in \{\boldsymbol{y} \in R^2 : y_1 \le 1\} \cap \{\boldsymbol{y} \in R^2 : y_2 \le 1\} \\
&= 1/2 \qquad \text{for} \quad \boldsymbol{x} \in \big(\{\boldsymbol{y} \in R^2 : 1 < y_1 \le 3\} \cap \{\boldsymbol{y} \in R^2 : 1 < y_2 \le 3\}\big) \cup \\
&\qquad \cup \big(\{\boldsymbol{y} \in R^2 : 1 < y_1 \le 3\} \cap \{\boldsymbol{y} \in R^2 : 1 < y_2 \le 1\}\big) \cup \\
&\qquad \cup \big(\{\boldsymbol{y} \in R^2 : 1 < y_2 \le 3\} \cap \{\boldsymbol{y} \in R^2 : 1 < y_1 \le 1\}\big), \\
&= 0 \qquad \text{for} \quad \boldsymbol{x} \in \{\boldsymbol{y} \in R^2 : y_1 > 3\} \cup \{\boldsymbol{y} \in R^2 : y_2 > 3\}.
\end{aligned}
$$

The structure of $\kappa_{\boldsymbol{C}}(\boldsymbol{x}) = \max(\kappa_{\mathcal{K}}(\boldsymbol{x}), \kappa_{\mathcal{L}}(\boldsymbol{x}))$ is not as complicated as the one of $\lambda_{\boldsymbol{C}}$. Namely, $\kappa_{\boldsymbol{C}}(\boldsymbol{x}) = \kappa_{\mathcal{K}}(\boldsymbol{x}) = \kappa_I(\boldsymbol{x})$ for all $\boldsymbol{x} \in R^2$. It means that for $\gamma_{\boldsymbol{C}}(\boldsymbol{x}) = \min(\kappa_{\boldsymbol{C}}(\boldsymbol{x}), \lambda_{\boldsymbol{C}}(\boldsymbol{x}))$ of the fuzzy core $\boldsymbol{C}_F$

$$
\begin{aligned}
\gamma_{\boldsymbol{C}}(\boldsymbol{x}) &= 1/2 \quad \text{for} \quad \boldsymbol{x} \in \{\boldsymbol{y} \in R^2 : 5 \le y_1 + y_2 \le 6\} \cap \\
&\qquad \cap \{\boldsymbol{y} \in R^2 : y_1 \ge 1\} \cap \\
&\qquad \cap \{\boldsymbol{y} \in R^2 : y_2 \ge 1\}, \ \boldsymbol{x} \notin \{(4,1),(1,4)\}, \\
&= 1 \quad \text{for} \quad \boldsymbol{x} \in \{(4,1),(1,4)\}, \\
&= 0 \quad \text{otherwise.}
\end{aligned}
$$

We see that $\gamma_{\boldsymbol{C}}(I, \mathcal{W}) = 1$, $\gamma_{\boldsymbol{C}}(\boldsymbol{x}) > 0$ for $\boldsymbol{x} \in C$, and that also $\eta(\mathcal{K}) = 1$, $\eta(\mathcal{L}) = 1$.

□

Part IV

ALTERNATIVE APPROACHES

The concepts and results shown in the previous two parts represent in some sense standard approaches to the fuzzification of the bargaining and coalition forming model in two main types of coalition games. Nevertheless, some components of fuzzy cooperation models presented above admit also a certain variability of their construction. In the following chapters we will briefly show at least some of the possible alternative approaches to the concepts investigated in the previous parts.

In some cases this variability of views reflects specific properties of fuzzy cooperation which have no remarkable roots in the deterministic theory. It concerns, namely, the comparison and certain disproportion between the models with and without side-payments observable in the games which we call here "linear," and the modification of the domination concept.

Another alternative to the above models which is mentioned in this part, namely a modified motivation to cooperation and considerations about relation between strategic and coalition games, has its counterpart also in the deterministic coalition game theory and here it is only briefly recalled and discussed from the fuzzy set theoretical point of view.

One of the following chapters deals with an alternative approach to the domination relation and its consequences for the form of superoptima and Pareto optima.

The last chapter briefly describes an alternative approach to the processing of fuzzy quantities which represent some vague verbal expressions. It appears to be an alternative to Chapter 2 and offers interesting tools for modification (and, probably, simplification) of procedures used in Part II.

13. Linear Coalition Games

In this chapter, we combine the properties of fuzzy coalition games with and without side-payments. It implies certain demands regarding the notations and a careful distinguishing of sophisticated differences between similar (but not identical) symbols used for the two types of games.

We will use the term "*linear coalition game*" for a link between coalition games with and without side-payments. They represent a class of coalition games in which both models of cooperation can be compared. Namely, every deterministic coalition game with side-payments (I, v) can be obviously considered as a special type of (deterministic) coalition game without side payments $(I, \mathcal{V})$ where for any coalition $K \subset I$

$$\mathcal{V}(K) = \left\{ x \in R^I : \sum\nolimits_K x_i \leq v(K) \right\}. \tag{13.1}$$

It is also not difficult to verify that many concepts introduced for (I, v) and $(I, \mathcal{V})$ in specific ways typical for respective types of games, like the superadditivity, subadditivity, additivity, core or convexity, do in fact coincide.

For example, the inclusion

$$\mathcal{V}(K \cup L) \supset \mathcal{V}(K) \cap \mathcal{V}(L)$$

for disjoint $K, L \subset I$ is equivalent to the condition

$$v(K \cup L) \geq v(K) + v(L)$$

(cf. [21]).

The core defined by (3.6)

$$C = \left\{ x \in R^I : \sum\nolimits_I x_i \leq v(I), \forall\, K \subset I, \sum\nolimits_K x_i \geq v(K) \right\}$$

is identical with the set (3.21)

$$\boldsymbol{C} = \mathcal{V}(I) \cap \left(\bigcap\nolimits_{K \subset I} \mathcal{V}^*(K) \right).$$

The Pareto optima for $K \subset I$,

$$\left\{x \in R^I : \sum_K x_i = v(K)\right\} \quad \text{and} \quad \mathcal{V}(K) \cap \mathcal{V}^*(K)$$

represent identical sets of imputations.

In this chapter we illustrate on a few concepts that in the fuzzy coalition games such a strong link does not generally exist. This fact follows from the disproportions between the fuzzifications of both methodologies – the one based on operations with numbers $v(K)$ and the one based on set theoretical operations with sets $\mathcal{V}(K)$.

First of all, we have to formulate a link between both basic types of games, analogous to (13.1), for the fuzzy extensions of (I, v) and $(I, \mathcal{V})$. It is obvious that even if the deterministic games fulfill (13.1), the fuzzy extension $(I, \mathcal{W})$ of $(I, \mathcal{V})$ can be generally very far from any fuzzy extension (I, w) of (I, v).

Let (I, v) and $(I, \mathcal{V})$ be deterministic coalition games fulfilling (13.1). Let (I, w) be an arbitrary fuzzy extension of (I, v). Let us construct the fuzzy extension $(I, \mathcal{W})$ of $(I, \mathcal{V})$ such that for any coalition K

$$\begin{aligned} \kappa_K(x) &= \mu_K(y) && \text{for} \quad x \in R^I,\ \textstyle\sum_K x_i = y > v(K), \\ &= 1 && \text{for} \quad x \in R^I,\ \textstyle\sum_K x_i = y \leq v(K), \end{aligned} \tag{13.2}$$

where, as above, μ_K is the membership function of the fuzzy quantity $w(K) \in \mathbb{R}$ and κ_K is the membership function of fuzzy set $\mathcal{W}(K) \subset R^I$.

Then, we say that $(I, \mathcal{W})$ is a *linear fuzzy coalition game* extending (I, v) and $(I, \mathcal{V})$.

Definition (13.2) can be substituted by an equivalent expression

$$\kappa_K(x) = \sup\left(\mu_K(y) : y \geq \sum_K x_i\right).$$

This means that the crisp coalition game with side-payments (I, v), in fact, has two fuzzy extensions, namely (I, w) and $(I, \mathcal{W})$ where the second one is also the fuzzy extension of $(I, \mathcal{V})$ fulfilling (13.1).

Some properties in which $(I, \mathcal{W})$ differs from (I, w) illustrate the basic differences and similarities between the fuzzy numerical and fuzzy set theoretical approach to the vagueness in cooperative behaviour.

In this brief note we touch two groups of properties, the ones regarding the superadditivity and the core.

The superadditivity of the fuzzy coalition game (I, w) is defined by means of (5.2) and (5.4) in Subsection 5.1 and it is a fuzzy property being valid with possibility $\nu_{\text{super}}(I, w)$.

On the other hand, the superadditivity of $(I, \mathcal{W})$ defined by (11.1) is a deterministic property which can be additionally fuzzified by means of the procedure described in Subsection 11.2 and the possibility of superadditivity is denoted by $\nu_{\text{super}}(I, \mathcal{W})$.

The results summarized in the subsections mentioned here imply that $(I, \mathcal{W})$ is superadditive in the deterministic imputational sense iff it is fuzzy superadditive with possibility $\nu_{\text{super}}(I, \mathcal{W}) = 1$ which implies that the deterministic game $(I, \mathcal{V})$ is superadditive in the sense of (3.12).

This is equivalent to the superadditivity of (I, w) as shown in [20] and implies that (I, w) is superadditive with possibility $\nu_{\text{super}}(I, w) = 1$. The opposite implications are not valid as shown by the following example.

Example 13.1 Let us consider 2-players coalition game with side-payments (I, v) such that $I = \{1, 2\}$, $v(\{i\}) = 1$, $i = 1, 2$, $v(I) = 3$. It means that we have also introduced a coalition game without side-payments $(I, \mathcal{V})$, where

$$\mathcal{V}(\{i\}) = \{\boldsymbol{x} \in R^2 : x_i \leq 1\}, \quad i = 1, 2, \qquad \mathcal{V}(I) = \{\boldsymbol{x} \in R^2 : x_1 + x_2 \leq 3\}.$$

Both games are, evidently, superadditive in the proper sense.

Let us consider now their fuzzy extensions (I, w) and $(I, \mathcal{W})$ such that for $i = 1, 2$, $x \in R$, $\boldsymbol{x} = (x_1, x_2) \in R^2$:

$$\begin{aligned} \mu_{\{i\}}(x) &= x & &\text{for} & &x \in [0, 1], \\ &= (5 - x)/4 & &\text{for} & &x \in [1, 5], \\ &= 0 & &\text{otherwise,} \end{aligned}$$

$$\begin{aligned} \mu_I(x) &= x - 2 & &\text{for} & &x \in [2, 3], \\ &= 4 - x & &\text{for} & &x \in [3, 4], \\ &= 0 & &\text{otherwise,} \end{aligned}$$

$$\begin{aligned} \kappa_{\{i\}}(\boldsymbol{x}) &= 1 & &\text{for} & &x_i \leq 1, \\ &= (5 - x_i)/4 & &\text{for} & &x_i \in [1, 5], \\ &= 0 & &\text{for} & &x_i > 5, \end{aligned}$$

$$\begin{aligned} \kappa_I(\boldsymbol{x}) &= 1 & &\text{for} & &x_1 + x_2 \leq 3, \\ &= 4 - (x_1 + x_2) & &\text{for} & &3 < x_1 + x_2 < 4, \\ &= 0 & &\text{for} & &x_1 + x_2 \geq 4. \end{aligned}$$

Then, obviously,

$$\nu_{\succsim}(w(I), w(\{1\}) \oplus w(\{2\})) = 1$$

and, consequently, $\nu_{\text{super}}(I, w) = 1$.

On the other hand, $\mathcal{W}(\{1\}) \cap \mathcal{W}(\{2\})$ is not a subset of $\mathcal{W}(I)$ and also for any α-cut, $\alpha > 0$, $\mathcal{W}_\alpha(\{1\}) \cap \mathcal{W}_\alpha(\{2\})$ is not subset of $\mathcal{W}_\alpha(I)$.

□

The previous result can be extended for a more general fuzzy superadditivity. Let us suppose that $(I, \mathcal{W})$ is fuzzy superadditive with possibility $\nu_{\text{super}}(I, \mathcal{W})$ defined by (11.5). Due to this formula and due to (13.2), for any pair of disjoint coalitions $K, L \subset I$, if

$$\mathcal{W}(K \cup L) \supset_\alpha \mathcal{W}(K) \cap \mathcal{W}(L),$$

which means, for $\alpha \in [0,1]$,

$$\kappa_{K \cup L}^{(\alpha)}(\boldsymbol{x}) \geq \min\left(\kappa_K^{(\alpha)}(\boldsymbol{x}), \kappa_L^{(\alpha)}(\boldsymbol{x})\right),$$

then also

$$\mu_{K \cup L}(\boldsymbol{x}) \geq \min(\mu_K(\boldsymbol{x}), \mu_L(\boldsymbol{x}))$$

if the referred minimum is not greater than α.

This means that the possibility

$$\nu_{\succsim}(w(K \cup L), (w(K) \oplus w(L)))$$

of relation $w(K \cup L) \succsim w(K) \oplus w(L)$ is equal at least to α and, consequently, also

$$\nu_{\text{super}}(I, w) \geq \alpha.$$

The opposite implication is not true as shows the previous example (for $\alpha = 1$) as well as the next one.

Example 13.2 Let us consider a coalition game with side-payments (I, v) such that $I = \{1, 2\}$, $v = (\{i\}) = 2$, $i = 1, 2$, $v(I) = 3$. Its linear counterpart without side payments is then $(I, \mathcal{V})$ such that

$$\mathcal{V}(\{i\}) = \{\boldsymbol{x} \in R^2 : x_i \leq 2\}, \qquad \mathcal{V}(I) = \{\boldsymbol{x} \in R^2 : x_1 + x_2 \leq 3\}.$$

It is evident that none of these two games is superadditive in the proper sense. Let us consider, now, their fuzzy extensions (I, w) and $(I, \mathcal{W})$ such that for $x \in R$, $\boldsymbol{x} = (x_1, x_2) \in R^2$, $i = 1, 2$

$$\begin{aligned} \mu_{\{i\}}(x) &= x - 1 && \text{for} \quad x \in [1,2], \\ &= (5 - x)/3 && \text{for} \quad x \in [2,5], \\ &= 0 && \text{for} \quad x \notin [1,5], \end{aligned}$$

$$\begin{aligned} \mu_I(x) &= x - 2 && \text{for} \quad 2 \leq x_1 + x_2 \leq 3, \\ &= 4 - x && \text{for} \quad 3 < x_1 + x_2 \leq 4, \\ &= 0 && \text{otherwise}, \end{aligned}$$

$$\begin{aligned} \kappa_{\{i\}}(\boldsymbol{x}) &= 1 && \text{for} \quad x_i \leq 2, \\ &= (5 - x_i)/3 && \text{for} \quad 2 < x_i < 5, \\ &= 0 && \text{otherwise}, \end{aligned}$$

$$\begin{aligned} \kappa_I(\boldsymbol{x}) &= 1 && \text{for} \quad x_1 + x_2 \leq 3, \\ &= 4 - (x_1 + x_2) && \text{for} \quad 3 < x_1 + x_2 < 4, \\ &= 0 && \text{otherwise}. \end{aligned}$$

Then it is easy to verify that the membership function $\mu_{\{1\},\{2\}}$ of the sum $w(\{1\}) \oplus w(\{2\})$ is

$$\begin{array}{rclll} \mu_{\{1\},\{2\}}(x) & = & \frac{x}{2}-1 & \text{for} & x \in [2,4], \\ & = & (10-x)/6 & \text{for} & x \in (4,10], \\ & = & 0 & \text{otherwise,} & \end{array}$$

and, consequently, the possibility $\nu_{\succsim}(w(I), w(\{1\}) \oplus w(\{2\}))$ of the relation $w(I) \succsim w(\{1\})$
$\oplus(\{2\})$ is equal to $2/3$.

However, neither for $\alpha = 2/3$ nor for any other $\alpha > 0$ the inclusion $\mathcal{W}_\alpha(I) \supset W_\alpha(\{1\}) \cap \mathcal{W}_\alpha(\{2\})$ holds.

□

A similar situation appears also if the fuzzy cores of both games, (I, w) and $(I, \mathcal{W})$, are considered. The fuzzy core C_F of (I, w), as well as the fuzzy core $\boldsymbol{C}_F$ of $(I, \mathcal{W})$, are fuzzy subsets of R^I. Their membership functions γ_C and $\gamma_{\boldsymbol{C}}$ are defined by (6.1) and (12.3), respectively. It can be easily seen (cf. [21]) that in the deterministic case the cores of (I, v) and $(I, \mathcal{V})$ are identical if (13.1) is fulfilled for all coalitions.

If the fuzzy extensions (I, w) and $(I, \mathcal{W})$ fulfilling (13.2) are considered, then the results presented in the previous chapters imply the following simple relation. Namely, if $\boldsymbol{x}$ belongs to the core C of (I, w), then $\gamma_C(\boldsymbol{x}) = 1$ and $\gamma_{\boldsymbol{C}}(\boldsymbol{x}) \in (0, 1]$.

This means that, for such an $\boldsymbol{x}$,

$$\gamma_C(\boldsymbol{x}) \geq \gamma_{\boldsymbol{C}}(\boldsymbol{x}) > 0. \tag{13.3}$$

For any $\boldsymbol{x} \in R^I$, the relations between the fuzzy cores C_F and $\boldsymbol{C}_F$ is as follows. Let us consider, first, the "strictly monotone" games, i.e. the fuzzy extension (I, w) of (I, v) such that for all coalitions $K \in 2^I$, all $y \in R$ such that $y > v(K)$ and $\mu_K(y) > 0$, and all $x \in R$, such that $v(K) \geq x > y$, the inequality $\mu_K(x) > \mu_K(y)$ holds.

Then, due to (13.2), for any $K \in 2^I$,

$$\kappa_K(\boldsymbol{x}) > \kappa_K(\boldsymbol{y})$$

for all $\boldsymbol{x}, \boldsymbol{y} \in R^I$ such that $\boldsymbol{x} \operatorname{dom}_K \boldsymbol{y}$, and

$$v(K) \geq \sum\nolimits_K x_i, \quad \kappa_K(\boldsymbol{y}) > 0.$$

So, in accordance with Chapter 10, the fuzzy superoptimum $\mathcal{W}^*(K)$ is identical to the deterministic superoptimum

$$\mathcal{V}(K) = \left\{\boldsymbol{x} \in R^I : \sum\nolimits_K x_i \geq v(K)\right\},$$

i.e.

$$\begin{aligned}\kappa_K^*(\boldsymbol{x}) &= 0 \quad \text{for } \boldsymbol{x} \in R^I \text{ such that } \textstyle\sum_K x_i < v(K),\\ &= 1 \quad \text{for } \boldsymbol{x} \in R^I \text{ such that } \textstyle\sum_K x_i \geq v(K).\end{aligned}$$

Let us compare the formulas for the fuzzy cores C_F and $\boldsymbol{C}_F$. For C_F of (I, w), the membership function $\gamma_C(\boldsymbol{x})$ is given by (6.1) as

$$\gamma_C(\boldsymbol{x}) = \min(\kappa_C(\boldsymbol{x}), \lambda_C(\boldsymbol{x})),$$

and for $\boldsymbol{C}_F$ of $(I, \mathcal{W})$ its membership function $\gamma_{\boldsymbol{C}}(\boldsymbol{x})$ is given by (12.3) as

$$\gamma_{\boldsymbol{C}}(\boldsymbol{x}) = \min(\kappa_{\boldsymbol{C}}(\boldsymbol{x}), \lambda_{\boldsymbol{C}}(\boldsymbol{x})).$$

and notice a subtle differences in the symbols that are significant in this chapter.

Relation (13.2) immediately implies that

$$\kappa_C(\boldsymbol{x}) = \kappa_{\boldsymbol{C}}(\boldsymbol{x}) \quad \text{for all } \boldsymbol{x} \in R^I \tag{13.4}$$

as for any coalition $K \subset I$ and any $\boldsymbol{x} \in R^I$

$$\begin{aligned}\nu_{\succsim}(w(K), \textstyle\sum_K x_i) &= \kappa_K(\boldsymbol{x}) = 1 && \text{for } \textstyle\sum_K x_i \leq v(K),\\ &= \mu_K\left(\textstyle\sum_K x_i\right) = \kappa_K(\boldsymbol{x}) && \text{for } \textstyle\sum_K x_i > v(K).\end{aligned}$$

If $\boldsymbol{x} \in \boldsymbol{C}$ then for each $K \subset I$

$$\sum_K x_i \geq v(K) \quad \text{and} \quad \kappa_K^*(\boldsymbol{x}) = 1$$

and, consequently,

$$\lambda_C(\boldsymbol{x}) = 1 = \lambda_{\boldsymbol{C}}(\boldsymbol{x}). \tag{13.5}$$

If $\boldsymbol{x} \notin \boldsymbol{C}$, and there does exist a coalition structure $\mathcal{K}$ such that for all $K \in \mathcal{K}$

$$\sum_K x_i \leq v(K)$$

then there exists $L \subset I$ for which

$$\sum_L x_i < v(L)$$

and then $\kappa_L^*(\boldsymbol{x}) = 0$ which implies $\lambda_{\boldsymbol{C}}(\boldsymbol{x}) = 0$ but, generally,

$$\nu_{\succsim}\left(\sum_L x_i, w(L)\right) \in [0,1]$$

and, consequently, $\lambda_C(\boldsymbol{x}) \geq \lambda_{\boldsymbol{C}}(\boldsymbol{x})$.

If $\boldsymbol{x} \notin \boldsymbol{C}$ and for all $L \subset I$,

$$\sum_L x_i \geq v(L)$$

then, for all $L \subset I$,

$$\nu_{\succsim}\left(\sum_{L} x_i, w(L)\right) = 1 = \kappa_L^*(\boldsymbol{x})$$

which means that

$$\lambda_C(\boldsymbol{x}) = \lambda_{\boldsymbol{C}}(\boldsymbol{x}) = 1$$

and $\gamma_C(\boldsymbol{x}) = \kappa_C(\boldsymbol{x}) = \kappa_{\boldsymbol{C}}(\boldsymbol{x}) = \gamma_{\boldsymbol{C}}(\boldsymbol{x})$.

The previous relations mean that, generally,

$$\gamma_C(\boldsymbol{x}) \geq \gamma_{\boldsymbol{C}}(\boldsymbol{x})$$

for the games of the considered type.

The procedure for games in which the membership functions are not "strictly monotone" in the above sense is in some steps analogous to the previous one. Equality (13.4) remains fulfilled for all $\boldsymbol{x} \in R^I$, and also, for $\boldsymbol{x} \in \boldsymbol{C}$ the general relation (13.3) is true.

If $\boldsymbol{x} \in R^I$ is such that for all $K \in 2^I$

$$\sum_{K} x_i > v(K)$$

then, evidently, there does not exist any coalition structure $\mathcal{K}$ for which

$$\boldsymbol{x} \in \bigcap_{K \in \mathcal{K}} \mathcal{V}(K).$$

This means that $\kappa_C(\boldsymbol{x}) = \kappa_{\boldsymbol{C}}(\boldsymbol{x})$ can, generally, take on a value from the interval $[0, 1)$, due to the actual form of the membership functions μ_K.

On the other hand, such an $\boldsymbol{x}$ fulfills

$$\sum_{L} x_i > v(L) \quad \text{for all} \quad L \subset I,\ L \neq \emptyset.$$

This means that the possibility of

$$\sum_{L} x_i \succsim w(L)$$

is equal to

$$\nu_{\succsim}\left(\sum_{L} x_i, w(L)\right) = 1 \quad \text{for all } L \subset I,$$

and, consequently, $\lambda_C(\boldsymbol{x}) - 1$, and $\gamma_C(\boldsymbol{x}) = \kappa_C(\boldsymbol{x})$.

As the fuzzy games (I, w) and $(I, \mathcal{W})$ are not "strictly monotone," the values of membership function $\kappa_L^*(\boldsymbol{x})$ are, generally, in the interval $(0, 1]$, and also $\lambda_{\boldsymbol{C}}(\boldsymbol{x})$ can be a value from the same interval.

Regarding these results,

$$\gamma_C(\boldsymbol{x}) = \min(\lambda_C(\boldsymbol{x}), \kappa_C(\boldsymbol{x})) \geq \min(\lambda_{\boldsymbol{C}}(\boldsymbol{x}), \kappa_{\boldsymbol{C}}(\boldsymbol{x})) = \gamma_{\boldsymbol{C}}(\boldsymbol{x}).$$

We have shown that for the linear coalition games fulfilling (13.1),(13.2) is generally

$$\gamma_C(\boldsymbol{x}) \geq \gamma_{\boldsymbol{C}}(\boldsymbol{x}).$$

This result is analogously valid for all $\boldsymbol{x} \in R^I$ such that in any coalition structure $\mathcal{K}$ there exists a coalition $K \in \mathcal{K}$ for which

$$\sum_K x_i > v(K).$$

This means the inclusion $C_F \supset \boldsymbol{C}_F$ of fuzzy cores. A natural interpretation of this inclusion can compare the demands on the stability of solutions of the fuzzy extensions (I, w) and $(I, \mathcal{W})$ represented by fuzzy cores.

The fuzzy core of $(I, \mathcal{W})$ is more demanding than the fuzzy core of (I, w) even if both fuzzy games extend practically the same game and both fuzzy cores are fuzzifications of the same deterministic set of imputations $C = \boldsymbol{C}$.

The main conclusion following from this chapter shows the principal differences between the two approaches to the fuzziness in coalition games, formally represented by the fuzzy numerical model in (I, w) and the fuzzy set model in $(I, \mathcal{W})$.

To illustrate a relation between the two versions of the fuzzy coalition games, namely the ones with and without side-payments, it is useful to compare Example 6.1 with Example 12.1, and Example 6.2 with Example 12.2 (see also Figure 6 and Figure 12.1). Each of these pairs represents two views at one linear coalition game.

14. A Modified Model of Domination and Superoptimum

In this chapter, we briefly discuss quite a natural question regarding the domination concept in fuzzy coalition games without side-payments. The domination relation between two imputations via a coalition, for the deterministic case, was defined in Chapter 3 by formula (3.15).

If $\boldsymbol{x}, \boldsymbol{y} \in R^I$ and $K \subset I$, then the symbol $\boldsymbol{x} \operatorname{dom}_K \boldsymbol{y}$ denotes that $\boldsymbol{x}$ dominates $\boldsymbol{y}$ via K. An extension of the domination relation to fuzzy coalition games without side-payments was suggested in Chapter 10 as a fuzzy property which is valid with possibility $\nu_K(\boldsymbol{x}, \boldsymbol{y})$ given by (10.1). It is called a "significant domination" and its validity is conditioned by the deterministic domination (3.15).

Moreover, the possibility of the dominated imputation cannot be higher than the possibility of the dominating one. It makes sense to ask if the latter condition is not too strong and rather artificial, and if it can be simplified or completely ommitted.

In fact, it is evidently possible to substitute the concept of a "significant domination" by a simpler relation, let us call it a "*simple domination via a coalition*" which is a fuzzy ordering relation over R^I.

For every pair of imputations $\boldsymbol{x}, \boldsymbol{y} \in R^I$ and coalition K, we say that $\boldsymbol{x}$ *simply dominates* $\boldsymbol{y}$ via K iff $\boldsymbol{x} \operatorname{dom}_K \boldsymbol{y}$ in the sense of (3.15).

Moreover, we say that this relation is true with the possibility that the dominating imputation $\boldsymbol{x}$ is an element of the fuzzy set $\mathcal{W}(K)$, i.e. with the possibility $\kappa_K(\boldsymbol{x})$. The further condition, demanding $\kappa_K(\boldsymbol{x}) \geq \kappa_K(\boldsymbol{y})$, is omitted.

This approach to the domination appears quite natural. However, it can be interesting to check how it influences the form of the superoptimum sets.

Generally, for any coalition $K \subset I$ the superoptimum is a fuzzy subset of R^I containing all imputations in R^I which are not simply dominated via K by any imputation from $\mathcal{W}(K)$.

More formally, we denote this alternative fuzzy superoptimum by $\mathcal{W}^{**}(K)$, and

$$\mathcal{W}^{**}(K) = \left\{ \boldsymbol{y} \in R^I : \forall \boldsymbol{x} \in R^I, \boldsymbol{x} \operatorname{dom}_K \boldsymbol{y} \Rightarrow \boldsymbol{x} \notin \mathcal{W}(K) \right\}.$$

This means that $\mathcal{W}^{**}(K)$ is a fuzzy subset of R^I with the membership function $\kappa_K^{**} : R^I \to [0,1]$ such that for any $\boldsymbol{y} \in R^I$

$$\begin{aligned} \kappa_K^{**}(\boldsymbol{y}) &= \min\left(1 - \kappa_K(\boldsymbol{x}) : \boldsymbol{x} \in R^I, \boldsymbol{x} \operatorname{dom}_K \boldsymbol{y}\right) = \\ &= 1 - \max\left(\kappa_K(\boldsymbol{x}) : \boldsymbol{x} \in R^I, \boldsymbol{x}\operatorname{dom}_K \boldsymbol{y}\right). \end{aligned} \tag{14.1}$$

It is easy to see that, in general, $\kappa_K^*(\boldsymbol{x}) \geq \kappa_K^{**}(\boldsymbol{x})$, for all $\boldsymbol{x} \in R^I$. Let us focus our attention on the Pareto optima and coalitions with superoptima of the considered type.

It is not difficult to see that, generally (if the functions κ_K are continuous), the membership function of the intersection $\mathcal{W}(K) \cap \mathcal{W}^{**}(K)$, constructed as the minimum of κ_K and κ_K^{**}, fulfills the inequality

$$\min(\kappa_K(\boldsymbol{x}), \kappa_K^{**}(\boldsymbol{x})) \leq \frac{1}{2} \quad \text{for all } \boldsymbol{x} \in R^I.$$

The equality

$$\sup_{\boldsymbol{x} \in R^I} (\min(\kappa_K(\boldsymbol{x}), \kappa_K^{**}(\boldsymbol{x}))) = \frac{1}{2} \tag{14.2}$$

is guaranteed for a continuous κ_K.

If the continuity is not fulfilled, then the supremum (14.2) can also be greater than 1/2.

The core of a fuzzy coalition game $(I, \mathcal{W})$ is a subset of the Pareto optima of some coalitions. Namely, if $\mathcal{K}$ is the coalition structure which maximizes the value (12.1), then – for any $\boldsymbol{x} \in \boldsymbol{C}_F$ – both

$$\boldsymbol{x} \in \mathcal{W}(K) \quad \text{for all } K \in \mathcal{K}$$

and

$$\boldsymbol{x} \in \mathcal{W}^{**}(K) \quad \text{for all } K \in \mathcal{K}.$$

hold

This means that, if we use the modified fuzzy superoptimum $\mathcal{W}^{**}$ in the definition of a core [namely, in (12.2)], then usually (except for some cases with a discontinuous κ_K)

$$\sup(\gamma_{\boldsymbol{C}}(\boldsymbol{x}) : \boldsymbol{x} \in R^I) \leq \frac{1}{2}.$$

Hence, if we consider the possibility of using the alternative "simple domination" instead of the "significant domination", we have to take into account not only the simplicity or elegance of the basic definition but also the consequences for the Pareto optima and, especially, for the core. Namely, we have to decide if the core whose elements can belong to it at most with the possibility 1/2 is intuitively acceptable for our idea of a "strong" solution of the game.

The most important exception to the previous results is the degenerated case where for all $K \subset I$, $\kappa_K(\boldsymbol{x}) = 0$ for $\boldsymbol{x} \notin \mathcal{V}(K)$. Then, we also have $\gamma_{\boldsymbol{C}}(\boldsymbol{x}) = 1$ for $\boldsymbol{x} \in \boldsymbol{C}$, and $\gamma_{\boldsymbol{C}}(\boldsymbol{x}) = 0$ otherwise. But in this case we beglect the idea of fuzzification of the game at all.

The best method for the illustration of differences between both approaches to the domination relation is to compare the superoptima, eventually the Pareto optima, obtained by using the relations to the same games (or coalitions).

In the following examples we racall several coalitions already discused in Section 10. The presentation of a 1-player coalition offers an acceptable clearness of the results combined with a sufficient representability of the investigated shapes of the sets $\mathcal{W}(\{i\})$.

Example 14.1 Let us consider a coalition game $(I, \mathcal{V})$, its 1-player coalition $\{i\}$ such that $\mathcal{V}(\{i\}) = \{\boldsymbol{x} \in R^I : x_i \leq 2\}$, and its fuzzy extension $\mathcal{W}(\{i\})$ where

$$\begin{aligned} \kappa_{\{i\}}(\boldsymbol{x}) &= 1 && \text{for } x_i \leq 2, \\ &= 3 - x_i && \text{for } 2 < x_i < 3, \\ &= 0 && \text{for } x_i \geq 3 \end{aligned}$$

(compare with Example 10.1).

Then it is easy to see that

$$\begin{aligned} \kappa^{**}_{\{i\}}(\boldsymbol{x}) &= 0 && \text{for } x_i \leq 2, \\ &= x_i - 2 && \text{for } 2 < x_i < 3, \\ &= 1 && \text{for } x_i \geq 3, \end{aligned}$$

and for the Pareto optimum $\mathcal{Q}^*(\{i\}) = \mathcal{W}(\{i\}) \cap \mathcal{W}^{**}(\{i\})$ with the membership function $\rho^{**}_K : R^I \to [0, 1]$, there is

$$\begin{aligned} \kappa^{**}_{\{i\}}(\boldsymbol{x}) &= 0 && \text{for } x_i \leq 2 \text{ or } x_i \geq 3, \\ &= x_i - 2 && \text{for } 2 < x_i \leq 5/2, \\ &= 3 - x_i && \text{for } 5/3 < x_i < 3. \end{aligned}$$

The maximum value of $\rho^{**}_{\{i\}}$ is $\rho^{**}_{\{i\}}(\boldsymbol{x}) = \frac{1}{2}$ for $x_i = \frac{5}{2}$.

□

Example 14.2 Let us consider the coalition game $(I, \mathcal{V})$ already dealt with in the previous example and its 1-player coalition $\{i\}$ with $\mathcal{V}(\{i\})$ from Example 14.1 (and Examples 10.1 and 10.2).

Let us consider its fuzzy extension $\mathcal{W}(\{i\})$ with

$$\begin{aligned} \kappa_{\{i\}}(\boldsymbol{x}) &= 1 && \text{for } x_i \leq 3, \\ &= \tfrac{4}{x_i - 1} - 1 && \text{for } 3 < x_i < 5, \\ &= 0 && \text{for } x_i \geq 5, \end{aligned}$$

(compare with Example 10.2).

Then, $\mathcal{W}^{**}(\{i\})$ is defined by

$$\begin{aligned}\kappa^{**}_{\{i\}}(\boldsymbol{x}) &= 0 && \text{for} \quad x_i \leq 3, \\ &= 2 - \tfrac{4}{x_i - 1} && \text{for} \quad 3 < x_i < 5, \\ &= 1 && \text{for} \quad x_i \geq 3,\end{aligned}$$

and the Pareto optimum $\mathcal{Q}^*(\{i\}) = \mathcal{W}(\{i\}) \cap \mathcal{W}^{**}(\{i\})$ is given by membership function

$$\begin{aligned}\rho^{**}_{\{i\}}(\boldsymbol{x}) &= 0 && \text{for} \quad x_i \leq 3 \text{ or } x_i \geq 5, \\ &= \tfrac{4}{x_i - 1} - 1 && \text{for} \quad 3 < x_i \leq 11/3, \\ &= 2 - \tfrac{4}{x_i - 1} && \text{for} \quad 11/3 < x_i < 5.\end{aligned}$$

The maximal value of $\rho^{**}_{\{i\}}$ is $\rho^{**}_{\{i\}}(\boldsymbol{x}) = 1/2$ for $x_i = 11/3$.

□

Example 14.3 Even in this case we consider the game $(I, \mathcal{V})$ discussed in the previous two examples and its 1-players coalition $\{i\}$. In this case we define the fuzzy extension $\mathcal{W}(\{i\})$ of $\mathcal{V}(\{i\})$ analogously as in Example 10.3, i.e.

$$\begin{aligned}\kappa_{\{i\}}(\boldsymbol{x}) &= 1 && \text{for} \quad x_i \leq 2, \\ &= 1/2 && \text{for} \quad 2 < x_i < 3, \\ &= 0 && \text{for} \quad x_i > 3.\end{aligned}$$

Then, $\mathcal{W}^{**}(\{i\})$ is defined by

$$\begin{aligned}\kappa^{**}_{\{i\}}(\boldsymbol{x}) &= 0 && \text{for} \quad x_i < 2, \\ &= 1/2 && \text{for} \quad 2 \leq x_i < 3, \\ &= 1 && \text{for} \quad x_i \geq 3,\end{aligned}$$

the Pareto optimum $\mathcal{Q}^*(\{i\}) = \mathcal{W}(\{i\}) \cap \mathcal{W}^{**}(\{i\})$ is defined by

$$\begin{aligned}\rho^{**}_{\{i\}}(\boldsymbol{x}) &= 0 && \text{for} \quad x_i < 2 \text{ or } x_i > 3, \\ &= 1/2 && \text{for} \quad 2 \leq x_i \leq 3\end{aligned}$$

and also in this case

$$\max\left(\rho^{**}_{\{i\}}(\boldsymbol{x}) : \boldsymbol{x} \in R^I\right) = \frac{1}{2}.$$

□

In the following example we show a fuzzy extension of $\mathcal{V}(\{i\})$ for which $\rho^{**}_{\{i\}}(\boldsymbol{x}) > 1/2$ for some $\boldsymbol{x} \in R^I$.

Example 14.4 Let us consider a coalition game $(I, \mathcal{V})$, dealt with in the previous examples, its coalition $\{i\}$ and a fuzzy extension $\mathcal{W}(\{i\})$ of $\mathcal{V}(\{i\})$ such that

$$\begin{aligned} \kappa_{\{i\}}(\boldsymbol{x}) &= 1 \quad \text{for} \quad x_i \leq 2, \\ &= 2/3 \quad \text{for} \quad 2 < x_i \leq 3, \\ &= 1/3 \quad \text{for} \quad 3 < x_i \leq 4, \\ &= 0 \quad \text{for} \quad x_i > 4. \end{aligned}$$

Then, we can see that

$$\begin{aligned} \kappa^{**}_{\{i\}}(\boldsymbol{x}) &= 0 \quad \text{for} \quad x_i < 2, \\ &= 1/3 \quad \text{for} \quad 2 \leq x_i < 3, \\ &= 2/3 \quad \text{for} \quad 3 \leq x_i < 4, \\ &= 1 \quad \text{for} \quad x_i \geq 4. \end{aligned}$$

This means that the Pareto optimum $\mathcal{W}(\{i\}) \cap \mathcal{W}^*(\{i\})$ is defined by

$$\begin{aligned} \rho^{**}_{\{i\}}(\boldsymbol{x}) &= 0 \quad \text{for} \quad x_i < 2 \text{ or } x_i > 4, \\ &= 1/3 \quad \text{for} \quad 2 \leq x_i < 3 \text{ or } 3 < x_i \leq 4, \\ &= 2/3 \quad \text{for} \quad x_i = 3. \end{aligned}$$

This means that $\max(\rho^{**}_{\{i\}}(\boldsymbol{x}) : \boldsymbol{x} \in R^I) = 2/3 > 1/2$.

□

15. Strategic Background of Fuzzy Cooperation

The usual coalition game models assume that coalitional pay-offs are given among elementary input data of the game. Their source or the mechanism of their generation is usually not explicitly explained.

However, there exists a link between the basic game theoretical concept of strategies and the characteristic functions of coalition games. Let us mention, at least briefly, the fuzzified form of it.

We denote by I the set of players and for each player $i \in I$ we denote by S_i the set of his or her *strategies* (for this purpose the sets S_i can be sets of pure as well as mixed strategies). If $K \subset I$ is a coalition then we denote by S_K the Cartesian product

$$S_K = \times_{i \in K} S_i \tag{15.1}$$

and we will consider S_K for the set of strategies of the coalition K.

In the strategic games (cf. [44, 19]) every player $i \in I$ evaluates his or her income by means of a *utility function* , u_i, where

$$u_i : \times_{j \in I} S_j \to R,$$

and the value

$$u_i(\boldsymbol{s}), \quad \boldsymbol{s} = (s_j)_{j \in I} \in \times_{j \in I} S_j = S_I$$

specifies the utility obtained by player i if the players employ their strategies $s_j \in S_j$, $j \in I$.

For the coalition game without side-payment the general characteristic function $\mathcal{V}(K)$ of any coalition K is the set

$$\mathcal{V}(K) = \left\{ \boldsymbol{x} \in R^I : \exists\, \boldsymbol{s} = (s_i)_{i \in I} \in \times_{j \in I} S_j, \forall\, i \in K, x_i \leq u_i(\boldsymbol{s}) \right\}. \tag{15.2}$$

For the coalition game with side-payments, it is advantageous to define for every coalition $K \subset I$ a coalitional utility function u_K as the sum

$$u_K(\boldsymbol{s}) = \sum_{i \in K} u_i(\boldsymbol{s}), \quad \boldsymbol{s} \in \times_{j \in I} S_j. \tag{15.3}$$

Then, from the point of view of the coalition K, the original game turns into a 2-players strategic game with players K and $I-K$ (where the set $I-K$ can be empty which does not break the general validity of the following steps).

The strategic games can be approached in various ways. One of the standard solutions is a so-called guaranteed or minimax solution which is connected with the concept of a *value* of the game (cf. [44, 19] and many other works on strategic games).

Let us consider a 2-players game with players K and $I-K$. Their sets of strategies are S_K and S_{I-K}, and their utility functions are u_K and u_{I-K}.

So, it is possible to define a number

$$v_K = \max_{\boldsymbol{s}_K \in S_K} \min_{\boldsymbol{s}_{I-K} \in S_{I-K}} u_K(\boldsymbol{s}_K, \boldsymbol{s}_{I-K}), \tag{15.4}$$

where $\boldsymbol{s}_K = (s_i)_{i \in K}$, $\boldsymbol{s}_{I-K} = (s_i)_{i \in I-K}$, which represents a guaranteed income of player (coalition) K in the considered game. These values can be taken, in some cases, for the values of the characteristic function v in (I, v), i.e.

$$v(K) = v_K.$$

Of course, in some coalition games the values $v(K)$ do not reflect the guaranteed, and rather pessimistic, expectations of the coalitional profit. In some, very optimistic, cases the minimum in (15.4) can be substituted by the maximum, sometimes (and it will be the usual case), the values $v(K)$ are somewhere between the pessimistic and optimistic expectations of the total profit of the coalition K.

However, if the approach represented by (15.4) is used, then the values $v(K)$ are linked with the values of the 2-person strategic games played by "collective" players K and $I-K$.

Both of these links between the strategic and coalition games, represented by (15.2) and (15.4), can be extended to the games with vague expected profits. In the following paragraphs we consider a strategic game with the set of players I, with the class of sets of strategies S_i, $i \in I$, and with utility functions $\boldsymbol{u}_i$, $i \in I$.

Let us suppose that the utilities $\boldsymbol{u}_i(\boldsymbol{s})$, $\boldsymbol{s} = (s_i)_{i \in I}$, are not fully deterministic and that their values are known only vaguely. Then, each utility function $\boldsymbol{u}_i$ is a fuzzy function which means that all values $\boldsymbol{u}_i(\boldsymbol{s})$ are fuzzy quantities with the membership functions

$$\zeta_{i,\boldsymbol{s}} : R \to [0, 1].$$

Notice the difference in notations. By u_i, u_K, we denote the deterministic pointwise utility functions, while $\boldsymbol{u}_i, \boldsymbol{u}_K$ are their fuzzy counterparts, usually their fuzzy extensions.

Every coalition $K \subset I$ has its set of strategies S_K defined by (15.1), and its utility function $\boldsymbol{u}_K$, too.

Analogously to (15.3), for any $\boldsymbol{s} \in S_I$,

$$\boldsymbol{u}_K(\boldsymbol{s}) = \sum\nolimits_{i \in K}^{\oplus} \boldsymbol{u}_i(\boldsymbol{s}) \tag{15.5}$$

is also a fuzzy quantity defined as the sum of fuzzy quantities $\boldsymbol{u}_i(\boldsymbol{s})$ (cf. (2.3) and (7.7) for the specification of $\sum^{\oplus}$). By $\zeta_{K,\boldsymbol{s}} : R \to [0,1]$ we denote the membership function of $\boldsymbol{u}_K(\boldsymbol{s})$.

Fuzzy utility functions can generate fuzzy characteristic functions of fuzzy coalition games analogously to the deterministic case. The procedure is simpler in the case of fuzzy coalition games without side-payments.

For a coalition $K \subset I$, the fuzzy characteristic function $\mathcal{W}(K)$ with the membership function $\kappa_K : R^I \to [0,1]$ is defined by

$$\begin{aligned} \kappa_K(\boldsymbol{x}) &= \\ &= \sup[\min_{i \in K}(\zeta_{i,\boldsymbol{s}}(y_i)) : \\ &\quad \boldsymbol{s} = (s_j)_{j \in I} \in \times_{j \in I} S_j, \boldsymbol{y} \in R^I, \forall i \in K : y_i \geq x_i]. \end{aligned} \tag{15.6}$$

Let the deterministic strategic game characterized by I, $(S_i)_{i \in I}$, $(u_i)_{i \in I}$ generate the deterministic coalition game without side-payments $(I, \mathcal{V})$ by means of (15.2). Let the fuzzy utility functions $\boldsymbol{u}_i$, $i \in I$, be fuzzy extensions of u_i, $i \in I$, in the sense that for any $\boldsymbol{s} = (s_i)_{i \in I}$,

$$\zeta_{i,\boldsymbol{s}}(x) = 1 \quad \text{for } x = u_i(\boldsymbol{s}).$$

Then, the pair $(I, (\mathcal{W}(K))_{K \subset I})$ is a fuzzy extension of $(I, (\mathcal{V}(K))_{K \subset I})$ with properties specified in Chapter 10.

A fuzzy coalition game with side-payments can be generated from fuzzy utilities, too, similarly to the procedure formalized by (15.4). We have already defined total fuzzy utility functions $\boldsymbol{u}_K$, for all $K \subset I$, by (15.5). Let us try to suggest a formula which defines the membership function μ_K of $w(K)$ analogously to (15.4).

If $K \subset I$ and if we denote for arbitrary $\boldsymbol{s} = (s_i)_{i \in I}$ by $(\boldsymbol{s}_K, \boldsymbol{s}_{I-K})$ the partition of $\boldsymbol{s}$ into strategies controlled by K and $I - K$, respectively, then we put, for any $x \in R$,

$$\mu_K(x) = \zeta_{K,\boldsymbol{s}^*}(x) \tag{15.7}$$

for that $\boldsymbol{s}^* \in S_I$ for which

$$u_K(\boldsymbol{s}^*) = \max_{\boldsymbol{s}_K \in S_K} \min_{\boldsymbol{s}_{I-K} \in S_{I-K}} u_K(\boldsymbol{s}_K, \boldsymbol{s}_{I-K}).$$

The transition from strategic to coalition games with vague pay-offs can be illustrated by quite a smple example of a 2-players game.

Example 15.1 Let us consider a strategic game with 2 players, $I = \{1, 2\}$, with the sets of 2 strategies of each player

Table 15.1. Pay-offs in the game from Example 15.1

s	$s^a = (s_1^{(1)}, s_2^{(1)})$	$s^b = (s_1^{(1)}, s_2^{(2)})$	$s^c = (s_1^{(2)}, s_2^{(1)})$	$s^d = (s_1^{(2)}, s_2^{(2)})$
$u_1(s)$	3	0	5	2
$u_2(s)$	7	3	4	0
$u_I(s)$	10	3	9	2

$$S_1 = \left(s_1^{(1)}, s_1^{(2)}\right) = S_2 = \left(s_2^{(1)}, s_2^{(2)}\right).$$

Notice that there exist the following possible coalitions in such a game: $\{1\}$, $\{2\}$ and I.

Suppose that the individual pay-offs $u_i(s_1, s_2)$, $i = 1, 2$, $s_1 \in S_1$ and $s_2 \in S_2$ are given by the following Table 15.1 which includes also the total pay-offs of coalition I (of course, for coalitions $\{i\}$, $i = 1, 2$, $u_{\{i\}} = u_i$).

Such a game is in the game-theoretical terminology called a 2×2 bimatrix game.

Using (15.2), we can derive

$$\mathcal{V}(\{1\}) = \{\boldsymbol{x} \in R^2 : x_1 \leq 5\}, \quad \mathcal{V}(\{2\}) = \{\boldsymbol{x} \in R^2 : x_2 \leq 7\},$$
$$\mathcal{V}(I) = \{\boldsymbol{x} \in R^2 : x_1 \leq 3, x_2 \leq 7\} \cup \{\boldsymbol{x} \in R^2 : x_1 \leq 5, x_2 \leq 4\},$$

and using (15.4), we get

$$v(\{1\}) = 5, \quad v(\{2\}) = 4, \quad v(I) = 10.$$

Let us consider a fuzzy extension of the game $(I, (S_i), (u_i)_{i\in I})$ denoted $(I, (S_i)_{i\in I}, (\boldsymbol{u}_i)_{i\in I})$ where the fuzzy pay-offs $\boldsymbol{u}_i(\boldsymbol{s})$ are fuzzy quantities with the membership functions $\zeta_{i,\boldsymbol{s}}$ such that, for any $x \in R$ (compare with Table 15.1),

$$\begin{aligned}
\zeta_{1,\boldsymbol{s}^a}(x) &= 0 && \text{for} \quad x \leq 1,\\
&= (x-1)/2 && \text{for} \quad 1 < x \leq 3,\\
&= (5-x)/2 && \text{for} \quad 3 < x < 5,\\
&= 0 && \text{for} \quad x \geq 5,
\end{aligned}$$

$$\begin{aligned}
\xi_{2,\boldsymbol{s}^a}(x) &= 0 && \text{for} \quad x \leq 5,\\
&= (x-5)/2 && \text{for} \quad 5 < x \leq 7,\\
&= (9-x)/2 && \text{for} \quad 7 < x < 9,\\
&= 0 && \text{for} \quad x \geq 9,
\end{aligned}$$

and for $\boldsymbol{s} = \boldsymbol{s}^b, \boldsymbol{s}^c, \boldsymbol{s}^d$, $i = 1, 2$

$$\begin{aligned}\zeta_{i,s}(x) &= 0 && \text{for} \quad x \le u_i(s) - 1,\\ &= x - u_i(s) + 1 && \text{for} \quad u_i(s) - 1 < x \le u_i(s),\\ &= u_i(s) - x + 1 && \text{for} \quad u_i(s) < x < u_i(s) + 1,\\ &= 0 && \text{for} \quad x \ge u_i(s) + 1.\end{aligned}$$

Then, due to (15.5), for $u_I(s)$ the membership functions $\zeta_{I,s}$ are

$$\begin{aligned}\zeta_{I,s}(x) &= 0 && \text{for} \quad x \le 2 \text{ or } x \ge 10,\\ &= (x-2)/4 && \text{for} \quad 2 < x \le 6,\\ &= (10-x)/4 && \text{for} \quad 6 < x < 10,\end{aligned}$$

and for $s = s^b, s^c, s^d$,

$$\begin{aligned}\zeta_{I,s}(x) &= 0 && \text{for} \quad x \le u_1(s) + u_2(s) - 2,\\ &= (x - u_1(s) - u_2(s) + 2)/2 && \text{for} \quad u_1(s) + u_2(s) - 2 < x \le\\ &&& \qquad \le u_1(s) + u_2(s),\\ &= (u_1(s) + u_2(s) + 2 - x)/2 && \text{for} \quad u_1(s) + u_2(s) < x \le\\ &&& \qquad \le u_1(s) + u_2(s) + 2,\\ &= 0 && \text{for} \quad x > u_1(s) + u_2(s) + 2.\end{aligned}$$

Using (15.6), we can derive the membership functions of fuzzy sets $\mathcal{W}(\{1\}), \mathcal{W}(\{2\}), \mathcal{W}(I)$, for every $x \in R^2$, as

$$\begin{aligned}\kappa_{\{1\}}(x) &= 1 && \text{for} \quad x_1 \le 5,\\ &= 6 - x_1 && \text{for} \quad 5 \le x_1 < 6,\\ &= 0 && \text{for} \quad x_1 \ge 6,\end{aligned}$$

$$\begin{aligned}\kappa_{\{2\}}(x) &= 1 && \text{for} \quad x_2 \le 7,\\ &= (9 - x_2)/2 && \text{for} \quad 7 \le x_2 < 9,\\ &= 0 && \text{for} \quad x_2 \ge 9,\end{aligned}$$

$$\begin{aligned}\kappa_I(x) &= 1 && \text{for} \quad x \in \mathcal{V}(I),\\ &= 6 - x_1 && \text{for} \quad 5 < x_1 \le 6,\ x_2 \le x_1 - 1,\\ &= 5 - x_2 && \text{for} \quad 4 < x_1 \le 5,\ 2x_2 - 5 \le x_1 \le x_2 + 1,\\ &= (5 - x_1)/2 && \text{for} \quad 3 < x_1 \le 5,\ (5 - x_1)/2 \le x_2 \le x_1 + 4,\\ &= (9 - x_2)/2 && \text{for} \quad 7 < x_2 \le 9,\ x_1 \le x_2 - 4,\\ &= 0 && \qquad \text{otherwise.}\end{aligned}$$

In such a way we have defined a fuzzy coalition game without side-payments $(I, \mathcal{W})$ which is a fuzzy extension of the game $(I, \mathcal{V})$ with $\mathcal{V}(K)$ described above.

If we wish to construct a fuzzy coalition game with side-payments (I, w) which is a fuzzy extension of (I, v) with values $v(K)$ defined above, it is easy to use (15.7) where $s^* = s^c$ for the coalitions $\{1\}$ and $\{2\}$, and $s^* = s^a$ for the coalition I.

Then, we have

Table 15.2. Strategy vectors and individual pay-offs in Example 15.2

$s \in S_1 \times S_2$	$s^a = (s_1^{(1)}, s_2^{(1)})$	$s^b = (s_1^{(1)}, s_2^{(2)})$	$s^c = (s_1^{(2)}, s_2^{(1)})$	$s^d = (s_1^{(2)}, s_2^{(2)})$
$u_1(s)$	3	1	2	5
$u_2(s)$	3	2	1	5

$$\mu_{\{1\}}(x) = \zeta_{1,s^c}(x), \quad \mu_{\{2\}}(x) = \zeta_{2,s^c}(x), \quad \mu_I(x) = \zeta_{I,s^c}(x).$$

□

The procedure connected with the construction of the max-min solution like (15.7) can be generalized with respect to the fact that the ordering relation and the maximum and minimum concepts among fuzzy quantities are also fuzzy, valid with different possibilities for different elements [cf. (2.9),(2.14),(2.15)].

Obviously, even in a very elementary case of 2×2 strategies the procedure becomes combinatorially cumbersome. Its analog, for larger sets of strategies, probably goes beyond the scope of realistic applicability.

Just to illustrate the idea of a fuzzified approach to the minimax solution, we present the following example in which the input data are extremely simplified.

Example 15.2 Let us consider a 2-players strategic game $(\{1,2\}, (S_i)_{i=1,2}, (u_i)_{i=1,2})$ quite similar to that discussed in Example 15.1. In the following procedures we are interested in 1-pla yer coalitions $\{1\}$ and $\{2\}$. Evidently, there exist 4 coalition vectors from $S_1 \times S_2$ and Table 15.2 presents the deterministic pay-offs of both players. Analogously to the previous example we denote $S_1 = (s_1^{(1)}, s_1^{(2)})$, $S_2 = (s_2^{(1)}, s_2^{(2)})$.

The fuzzification of pay-offs $u_i(s)$ consists in their substitution by fuzzy quantities $\boldsymbol{u}_i(s)$ with the membership functions $\zeta_{i,s} : R \to [0,1]$ such that – for all $i = 1,2$, all $s \in S_1 \times S_2$ and all $x \in R$ – there holds

$$\begin{aligned} \zeta_{i,s}(x) &= (x - u_i(s) + 2)/2 \quad \text{for} \quad x \in [u_i(s) - 2, u_i(s)], \\ &= (u_i(s) + 2 - x)/2 \quad \text{for} \quad x \in [u_i(s), u_i(s) + 2], \\ &= 0 \quad \text{otherwise.} \end{aligned}$$

Let us focus our attention on the coalition $\{1\}$. In the first step we find for every fixed strategy $s_2 \in S_2$ of the anticoalition $\{2\}$ the possibility with which the pair of strategies (s_1, s_2) brings the minimal fuzzy pay-off $\boldsymbol{u}_1(s_1, s_2)$ from $\boldsymbol{u}_1(s_1^{(1)}, s_2)$ and $\boldsymbol{u}_1(s_1^{(2)}, s_2)$.

The ordering of fuzzy quantities and their fuzzy extremes wwas explained in Chapter 2, namely by means of the formulas (2.9),(2.15). First of all we have to find possibilities with which the relations $u_1(s_1^{(1)}, s_2) \succsim u_1(s_1^{(2)}, s_2)$ and $u_1(s_1^{(2)}, s_2) \succsim u_1(s_1^{(1)}, s_2)$ are valid for $s_2 = s_2^{(1)}, s_2^{(2)}$. Let us note that for any $s \in S_1 \times S_2$, $u_1(s) \succsim u_1(s)$ with possibility 1.

Then, it is easy to verify that

$$\begin{array}{lll} u_1(s^a) \succsim u_1(s^b) & \text{with possibility} & 1 \\ u_1(s^b) \succsim u_1(s^a) & \text{with possibility} & 1/2 \end{array}$$

and, consequently, $u_1(s^a)$ is the fuzzy minimum from $(u_1(s^a), u_1(s^b))$ with possibility

$$\nu_{\min}(u_1(s^a)) = 1/2$$

while

$$\nu_{\min}(u_1(s^b)) = 1.$$

Analogously, for the pair $u_1(s^c), u_1(s^d)$, we have

$$\begin{array}{lll} u_1(s^c) \succsim u_1(s^d) & \text{with possibility} & 1/4 \\ u_1(s^d) \succsim u_1(s^c) & \text{with possibility} & 1 \end{array}$$

and, simlarly as in the previous case, the fuzzy minimum from the pair $(u_1(s^c), u_1(s^d))$ is attained with the possibilities

$$\nu_{\min}(u_1(s^c)) = 1, \quad \nu_{\min}(u_1(s^d)) = 1/4.$$

In this way, we have constructed for every strategy $s_1 \in S_1$ of $\{1\}$ the fuzzy set of minimal pay-offs attainable for this strategy.

Let us continue our procedure. Now, we find the possibility with which the strategies $s_1^{(1)}$ and $s_1^{(2)}$ can become the "minimax" strategies of coalition $\{1\}$ in the sense analogous to the deterministic one (15.4). We start with the possibilities with which s^a, s^b, and s^c, s^d can act in the fuzzy minimax considerations.

The possibility that $s_1^{(1)}$ will be the minimax strategy is equal to the possibility that it will be such as a component of s^a or as a component of s^b. If we denote the possibilities by $\nu_{\text{minmax}}(s_1^{(1)})$, $\nu^*_{\text{minmax}}(s^a)$ and $\nu^*_{\text{minmax}}(s^b)$, respectively, then

$$\nu_{\text{minmax}}(s_1^{(1)}) = \max\left(\nu^*_{\text{minmax}}(s^a), \nu^*_{\text{minmax}}(s^b)\right),$$

where the right-hand-side possibilities can be derived as follows.

The possibility $\nu^*_{\text{minmax}}(s^a)$ that the strategy vector s^a is significant for the minimax strategy of the 1-player coalition $\{1\}$ is the possibility $\nu_{\min}(u_1(s^a))$ that s^a is minimal of (s^a, s^c) and that $u_1(s^c)$ is minimal of $(u_1(s^c), u_1(s^d))$ [i.e. $\nu_{\min}(u_1(s^c))$] and that $u_1(s^a) \succsim u_1(s^c)$, or it is the possibility that s^a is the minimal of (s^a, s^b) and s^d is the minimal of (s^c, s^d) and $u_1(s^a) \succsim u_1(s^d)$.

Summarizing these rather heuristic considerations in a more formal way, we see that

$$\begin{aligned} \nu^*_{\text{minmax}}(s^a) &= \max\Big[\big(\min(\nu_{\min}(s^a), \nu_{\min}(s^c), \nu_{\succsim}(u_1(s^a), u_1(s^c))), \\ &\qquad (\nu_{\min}(s^a), \nu_{\min}(s^d), \nu_{\succsim}(u_1(s^a), u_1(s^d)))\big)\Big] = \\ &= \max[\min(1/2, 1, 1/2), \min(1/2, 1/4, 1/2)] \;=\; 1/2. \end{aligned}$$

Quite analogously, we have

$$\begin{aligned} \nu^*_{\text{minmax}}(s^b) &= \max\Big[\big(\min(\nu_{\min}(s^b), \nu_{\min}(s^c), \nu_{\succsim}(u_1(s^b), u_1(s^c))), \\ &\qquad (\nu_{\min}(s^b), \nu_{\min}(s^d), \nu_{\succsim}(u_1(s^b), u_1(s^d)))\big)\Big] = \\ &= \max[\min(1, 1, 3/4), \min(1, 1/4, 0)] \;=\; 3/4. \end{aligned}$$

This means that, due to the above,

$$\nu_{\text{minmax}}(s_1^{(1)}) = \max(1/2, 3/4) = 3/4.$$

A similar procedure can be used when the strategy $s_1^{(2)}$ and its possibilities are analyzed. Repeating the above considerations, we can derive that

$$\nu_{\text{minmax}}\left(s_1^{(2)}\right) = \max\left(\nu^*_{\text{minmax}}(s^c), \nu^*_{\text{minmax}}(s^d)\right)$$

where

$$\begin{aligned} \nu^*_{\text{minmax}}(s^c) &= \max\Big[\big(\min(\nu_{\min}(s^c), \nu_{\min}(s^a), \nu_{\succsim}(u_1(s^c), u_1(s^a))), \\ &\qquad (\nu_{\min}(s^c), \nu_{\min}(s^b), \nu_{\succsim}(u_1(s^c), u_1(s^b)))\big)\Big] = \\ &= \max[\min(1, 1, 1), \min(1, 1/2, 3/4)] \;=\; 1, \end{aligned}$$

$$\begin{aligned} \nu^*_{\text{minmax}}(s^d) &= \max\Big[\big(\min(\nu_{\min}(s^d), \nu_{\min}(s^a), \nu_{\succsim}(u_1(s^d), u_1(s^a))), \\ &\qquad (\nu_{\min}(s^d), \nu_{\min}(s^b), \nu_{\succsim}(u_1(s^d), u_1(s^b)))\big)\Big] = \\ &= \max[\min(1/4, 1, 1), \min(1/4, 1/2, 1)] \;=\; 1/4, \end{aligned}$$

and, as a result,

$$\nu_{\text{minmax}}\left(s_1^{(2)}\right) = \max(1, 1/4) = 1.$$

Consequently, the 1-player coalition $\{1\}$ attains its minimax fuzzy pay-off with possibility 1 if it uses the strategy $s_1^{(2)}$, and it attains it with possibility 3/4 if it uses the strategy $s_1^{(1)}$.

The symmetry of values $u_1(\boldsymbol{s})$ and $u_2(\boldsymbol{s})$ in Table 15.2 means that the previous procedure can be repeated for the coalition $\{2\}$, by changing indices. Hence, the coalition $\{2\}$ attains its minimax fuzzy pay-off with possibility 1 if it uses the strategy $s_2^{(2)}$ and with possibility 3/4 if it uses the strategy $s_1^{(1)}$.

In the deterministic model we have denoted the minimax vector of strategies by $\boldsymbol{s}^*$ and it is easy to see that in our fuzzy model, $\boldsymbol{s}^d = \boldsymbol{s}^*$ with possibility

$$1 = \min\left(\nu_{\mathrm{minmax}}(s_1^{(2)}), \nu_{\mathrm{minmax}}(s_2^{(2)})\right)$$

while $\boldsymbol{s}^a$, $\boldsymbol{s}^b$ and $\boldsymbol{s}^c$ can become minimax vectors of strategies only with possibilities 3/4.

Let us note that in the original deterministic game formulated in the beginning of this example, $\boldsymbol{s}^* = (s_1^{(2)}, s_2^{(2)})$, as well.

□

16. Generation of Fuzzy Quantities

Not only the fuzzy coalition game theory but also a great majority of other fuzzy set theoretical models usually do not study the structure and origin of fuzzy data entering the model. Those data are simply taken for given. In some cases, especially if the data are in the verbal form ("approximately x", etc.), their structure is significant. They represent a certain kind of mixture of a quantitative and semantic component. One of them gives an idea about the numerical value, the second one informs about the structure of uncertainty connected with it. Each of them can be treated in a specific way.

This model of vague (fuzzy) data – in our case of values of the fuzzy characteristic function w in coalition games with side-payments – was not used in the previous chapters. However, we will mention it at least briefly to show one of possible alternative approaches to the entire model and its processing. The method is, in a more detailed way, described, e.g., in [38, 27, 31].

As we have already mentioned above, vague verbal descriptions of quantitative data have their quantitative and qualitative components as, e.g., "approximately 8," "very near to 10," "almost 5". This dualism exists, in quite a hidden form, even in such "anonymous" quantities like "many," "several," "few," etc. Due to the actual application, they can be interpreted like "very approximately x," "near to y," or "about z" where x, y, z correspond to the intuitively understood value following from the modelled situation.

Formally, each verbal quantity as, e.g., "approximately 8" is described by a *crisp numerical value* – in our case "8" – and a qualitative characterization of the type of vagueness – in our example "approximately."

This characterization is represented by a fuzzy quantity in a "normalized" form. Their membership functions are denoted by $\varphi : R \to [0,1]$ and called "*shapes*" of the verbal quantities. It is supposed that

$$\varphi(0) = 1 \qquad (16.1)$$

$$\varphi(x) \quad \text{is increasing for } x < 0 \text{ and decreasing for } x > 0. \qquad (16.2)$$

The symbols for shapes will be, in the following presentation, indexed by a verbal expression, which is represented by, e.g., $\varphi_{\mathbf{approximately}}$, and then, e.g.,

$$\varphi_{\text{approximately}}(x) = \max(0, x+1) \quad \text{for } x \leq 0,$$
$$= \max(0, 1-x) \quad \text{for } x \geq 0. \qquad (16.3)$$

The analytic form in (16.3) is rather variable and $\varphi_{\text{approximately}}$ can be, regarding the needs of a current application, defined by other functions too.

The membership function of a fuzzy quantity representing a verbal quantitative expression as, e.g., "approximately 8," is then denoted by $\mu_{\text{approximately 8}} : R \to [0,1]$ and

$$\mu_{\text{approximately8}}(x) = \varphi_{\text{approximately}}(x-8). \qquad (16.4)$$

The above procedure can be generalized in the following way. We can represent the source of a verbal data by a real valued function f, called *generator*, where

$$f(0) = 0, \qquad (16.5)$$
$$f(x) \quad \text{is continuous and strictly increasing.} \qquad (16.6)$$

Then, the membership functions are defined, instead of (16.4), by

$$\mu_{\text{approximately8}}(x) = \varphi_{\text{approximately}}(f(x) - f(8)). \qquad (16.7)$$

The first derivative of the generator f can measure the "reliability" or "exactness" of the source of verbal quantities. The higher it is the more concentrated or more exact data it offers. In formula (16.4), $f(x) = x$.

An example of the shape function φ can be

$$\varphi(x) = \max(0, 1 - |x|). \qquad (16.8)$$

In our example with "approximately 8" the combination of (16.8) and (16.7) implies

$$\mu_{\text{approximately8}}(x) = \max(0, 1 - |f(x) - f(8)|)$$

for some generating function f.

The main motivation for developing an alternative or, better to say, an analytic model of fuzzy quantities was implied by the idea to separate the quantitative and qualitative components of vague verbal quantity. Then, it is possible to process each of these components in a specific way, as this corresponds with an essential difference of their role in the description of a vague reality.

The quantitative component represented by "crisp values" can be processed by classic deterministic methods which lead to a deterministic "crisp core" of the result. The "shapes" (which are in some sense normalized membership functions of the fuzzy quantities describing the verbal descriptions of vagueness) can be dealt with by using procedures which are mostly based on fuzzy logic.

They include the traditional extension principle

$$\begin{aligned}
\varphi_c(x) &= \sup_{y\in R} (\min(\varphi_a(y), \varphi_b(x-y))) \quad \text{for addition } c = a+b, \\
\varphi_c(x) &= \sup_{y\in R, y\neq 0} (\min(\varphi_a(y), \varphi_b(x/y))) \quad \text{for multiplication } c = a\cdot b, \\
\varphi_c(x) &= \sup_{y,z\in R, x=F(y,z)} (\min(\varphi_a(y), \varphi_b(z))) \quad \text{generally, for } c = F(a,b),
\end{aligned}$$

but also fuzzy logical operations

$$\begin{aligned}
\varphi_c(x) &= \max(\varphi_a(x), \varphi_b(x)) \quad \text{for disjunction of } a \text{ and } b, \\
\varphi_c(x) &= \min(\varphi_a(x), \varphi_b(x)) \quad \text{for conjunction of } a \text{ and } b,
\end{aligned}$$

other combinations like

$$\begin{aligned}
\varphi_c(x) &= \varphi_a(x)\cdot\varphi_b(x), \\
\varphi_c(x) &= \varphi_a(x) + \varphi_b(x) - \varphi_a(x)\cdot\varphi_b(x),
\end{aligned}$$

or, more generally, a triangular norm T (see [41]),

$$\varphi_c(x) = T(\varphi_a(x), \varphi_b(x)).$$

Also the generators (16.5) and (16.6) can be combined. As their first derivatives reflect the exactness of the data source (cf. [27]), the combination can prevailably respect these derivatives.

As an output of this procedure we could obtain the resulting generator f and shape φ. Using a formula analogous to (16.7), we apply the resulting crisp value and derive the membership function of a fuzzy quantity corresponding to the vague output of the algorithm.

This briefly described method has quite general applications. In the above models we can use it for the computation of some quantities characterizing the fuzzy coalition games with side-payments. For example, the computation of Shapley values, discussed in Chapter 9, can be modified in the following way.

Let us consider a deterministic coalition game with side-payments (I, v) and its fuzzy extension (I, w). Let us suppose that for every coalition $K \subset I$, μ_K of $w(K)$ is increasing for $x < v(K)$ and decreasing for $x > v(K)$ and, of course, $\mu_K(v(K)) = 1$.

In the terms used above, every μ_K, $K \subset I$, is decomposable into a crisp core $v(K)$ and shape φ_K

$$\varphi_K(x) = \mu_K(x + v(K)). \tag{16.9}$$

Here we assume, for simplification, that all generators f_K, $K \subset I$, fulfill $f_K(x) = x$, for all $x \in R$. Then, we can use the deterministic formula (3.10) to compute crisp values t_i, $i \in I$, from the values $v(K)$, $K \subset I$. Moreover, we can combine the membership functions μ_K using some of the above formulas corresponding to the logical relations between fuzzy quantities $w(K)$. For example, for every $i \in I$, we can accept the hypothesis that the calculation of t_i and T_i needs the combination of $w(K)$ (it means, in fact, of μ_K or φ_K due to the extension principle and then we obtain fuzzy Shapley values T_i) by (9.1) based on (2.3) and (2.7). Or, we can accept another hypothesis, namely, that the calculation of a fuzzy Shapley value demands a logical conjunction of uncertainty connected with fuzzy quantities $w(K)$.

Then we can, for every player $i \in I$, combine the shapes φ_K, and derive φ_i, for example using

$$\varphi_i(x) = \min(\varphi_K(x) : K \subset I \text{ such that } i \in K), \quad x \in R. \tag{16.10}$$

We assume, quite naturally, that also the combined generators f_i derived from f_K, are linear and $f_i(x) = x$, for $x \in R$.

Then for any player $i \in I$ the membership function σ_i of his Shapley value (9.2) is not derived from (9.1) but

$$\sigma_i(x) = \varphi_i(x - t_i). \tag{16.11}$$

When the parameters (i.e. the values of characteristic function $v(K)$) are changed, it is easy to compute new values t_i, $i \in I$, and the shapes φ_i with respect to (16.9) and (16.10) (or another pattern of combination), and to find the fuzzy Shapley value. If, moreover, the structure of vagueness of the expected coalition pay-offs is unchanged, it means if only the values $v(K)$ alter but the shapes φ_K remain the same and

$$\mu_K(x) = \varphi_K(x - v(K)),$$

for all $K \subset I$, then the combined shapes φ_i remain also without any change and the membership functions σ_i can be computed from (16.11) only by the updating of the crisp cores t_i.

This alternative approach to the fuzzy coalition games seems to simplify their practical processing but its wider application depends on a further development of the general model of generators and shapes of fuzzy quantities.

The effect of this alternative approach on the analytic behaviour of the fuzzy coalition games without side-payments can be illustrated by the following example of processing one of the games already dealt with by means of more classical procedures in Example 7.2 and Example 7.4.

Example 16.1 Let us consider a 3-players game with side-payments (I, v) described as the game (I, w_{10}) in Example 7.2 and Example 7.4. It means that

$$v(\{i\}) = 1, \quad w(\{i,j\}) = 2, \quad v(I) = 2 \quad \text{for } i,j \in I,\ i \neq j.$$

Values $v(K)$ will be taken for crisp cores of the fuzzy quantities $w(K)$. Let us suppose that the vagueness included in their values is in all cases described by the shape functions

$$\begin{aligned}\varphi_K(x) &= (x+2)/2, && \text{for } x \in [-2,0]\\ &= (2-x)/2, && \text{for } x \in [0,2],\\ &= 0, && \text{for } x \notin [-2,2].\end{aligned}$$

Finally, we suppose that the reliability of the source of information about the vague values $w(K)$ is described by the scale function

$$\begin{aligned}f(x) &= 2x && \text{for } x \in [-10,10],\\ &= x+10 && \text{for } x \geq 10,\\ &= x-10 && \text{for } x \leq 5,\ x \leq -10.\end{aligned}$$

This means that for any coalition K the membership functions μ_K of $w(K)$ are given, with respect to

$$\mu_K(x) = \varphi_K(f(x) - f(v(K))).$$

$$\begin{aligned}\mu_K(x) &= x - v(K) + 1 && \text{for} && x \in [v(K)-1, v(K)],\\ &= v(K) - x + 1 && \text{for} && x \in [v(K), v(K)+1],\\ &= 0 && \text{otherwise.}\end{aligned}$$

In our case

$$\begin{aligned}\mu_{\{i\}}(x) &= x && \text{for } x \in [0,1],\\ &= 2-x && \text{for } x \in [1,2],\\ &= 0 && \text{for } x \notin [0,2],\end{aligned}$$

$$\begin{aligned}\mu_{\{i,j\}}(x) &= x-1 && \text{for } x \in [1,2],\\ &= 3-x && \text{for } x \in [2,3],\\ &= 0 && \text{for } x \notin [1,3],\end{aligned}$$

$$\mu_I(x) = \mu_{\{i,j\}}(x),\ x \in R.$$

We already know that the effectivity of I is attainable with possibility $\varepsilon(I) = 3/5$ (see Table 7.4). Let us suppose that the shapes are combined by means of the maximum operation, i.e. the shape function φ_{K+L} of $w(K) \oplus w(L)$ is defined by

$$\varphi_{K+L}(x) = \max(\varphi_K(x), \varphi_L(x))$$

as well as for φ_{K-L} of $w(K) \oplus (-w(L))$. In our case it means that $\varphi_{K+L}(x) = \varphi_{K-L}(x) = \varphi_K(x) = \varphi_L(x)$, the membership function μ_{K+L} of $w(K) \oplus w(L)$ is

$$\begin{aligned}\mu_{K+L}(x) &= x-(v(K)+v(L))+1 \quad \text{for } x\in[v(K)+v(L)-1,\ v(K)+v(L)],\\ &= v(K)+v(L)-x+1 \quad \text{for } x\in[v(K)+v(L),\ v(K)+v(L)+1],\\ &= 0 \quad \text{otherwise,}\end{aligned}$$

and the membership function $\mu_{K\cup L+K\cap L}$ of $w(K\cup L)\oplus w(K\cap L)$ is defined analogously.

Then, for $i,j,k\in I$, $i\neq j\neq k\neq i$,

$$\pi(\{i\},\{j\}) = \nu_{\succsim}(w(\{i,j\}),(w(\{i\})\oplus w(\{j\}))) = 1$$
$$\pi(\{i\},\{j,k\}) = 1/2, \quad \pi(\{i,j\},\{j,k\}) = 1/2.$$

It means that for the alternative processing of fuzzy quantities in this case

$$\varepsilon(I) \geq \pi^0(I,w)$$

cf. comments to Example 7.4.

□

Part V

CONCLUDING REMARKS

In the previous chapters we have presented basic and rather elementary steps of the development of an alternative model of cooperative behaviour under uncertainty. The uncertainty is not connected with the participation of players in particular coalitions but with the expectations of profit that can result from a coalitional cooperation.

Intuitively, it is possible to expect some kind of relation between both sources of uncertainty. For example, the degree of participation of a player at a particular coalition can be related to the expected profit resulting from this cooperation. This relation itself is not strict but essentially vague. Its formal structure has, intuitively, something in common with fuzziness and in this sense it corresponds with the nature of a general idea of a fuzzy cooperation. However, this has not been investigated either in this book or anywhere else. It remains an open question, and an open subject of a possible study – including the effectiveness of this topic.

The problems investigated in this book represent a specific view on vagueness (fuzziness) in cooperation and, consequently, the results presented here are the basic ones that are to be derived. They regard the very foundational concepts like the superadditivity, core and some other related ones.

Their formal structure is also somehow monotone – they show in what sense the fuzzified game represents an extension of the original deterministic game, in what degree of possibility the properties of the deterministic game are transferred to its fuzzy extension, and how this transfer reflects a formal shape of the membership functions of the input fuzzy pay-offs in both models: the one with side-payments, and the one without them.

There still exist many other game theoretical concepts which are related to the expected pay-offs (naturally, the expected pay-offs are the roots of other components of the game model). For example, we did not mention the von Neumann – Morgenstern solution or the nucleous or the kernel of the game. Their inclusion into the considered fuzzy coalitional game can clearly be interesting even if in some cases it can mean the need to introduce some auxiliary notions. So, for instance, the fuzzification of the von Neumann – Morgenstern solution would probably need the specification of a fuzzy imputation.

The results obtained in the previous chapters imply one general conclusion. Namely, the fuzzy extension of the input expectation of pay-offs extends the validity of some properties of the game (e.g., the extent of the Shapley value). On the other hand, some other properties, including even some important ones, valid in the deterministic case, are completely broken. Let us recall, e.g., the relation between the non-emptiness of core and the convexity or balancedness of a game with side-payments.

The properties of the fuzzified coalitional game extend the properties of its deterministic pattern in the case in which they are fulfilled for the deterministic case but their validity is not fulfilled (even with possibilities less than 1) if the deterministic game does not fulfil them.

We can conclude that the fuzzy coalition game has a much more complex structure than just a simple analogue of its deterministic counterpart. It has its own non-trivial properties and regularities that can be investigated independently of the properties of the deterministic coalitional games of both types.

Some of them, even if rather basic ones, have been indicated in this book. Namely, in the chapter on linear coalition games we can see that the fuzzification techniques used for both types of games with side-payments and without them are essentially different regarding the relation between, e.g., the fuzzy and deterministic core or fuzzy and deterministic superadditivity.

The case of linear coalition games shows that there exists a certain degree of freedom in the choice of the fuzzification method and that this choice influences the formal structure of the derived results.

This freedom is a special case of a more general flexibility of fuzzy set theoretical methods. Any application of the fuzzy set theoretical tools is connected with a more or less arbitrary choice of some elements exemplified by exact membership functions on the elementary level, and some methods of their processing in more advanced steps, like the orderings of fuzzy quantities, etc.

The fuzzy coalitional game model offers also its specific opportunities to find the flexibility mentioned above. In the previous chapters we have discussed an adequate relation of domination between imputations via a coalition, we have already mentioned a rich choice of definitions of ordering relation between fuzzy quantities, and we have already seen that there exist (at least) two possibilities of how to define the superadditivity of the fuzzy coalitional game without side-payments.

To discuss some other concepts mentioned in the above chapters, we would probably be able to find other possibilities of alternative approaches to their definitions. All of them rather limit the universal validity of the results and ideas presented in this book. As it is typical for any application of fuzzy set theory, there objectively exists a certain mixture of mathematical exactness in formal models and subjectivity in their choice. This is one of the properties that make the fuzzy set theoretical methods so attractive – their variability offers a choice of a constellation of tools adequate to specific features of each modelled practical situation.

References

[1]J. M. Adams: Fuzzy decision trees. Fuzzy Sets and Systems 4 (1980), 207–219.

[2]K. J. Arrow: Social Choice and Individual Values. Wiley, New York 1951.

[3]K. Asai, H. Tanaka, T. Okuda: Decision–making and its goal in a fuzzy environment. In: Fuzzy Sets and Their Applications for Cognitive and Decision Processes (L. A. Zadeh, Ed.). Academic Press, New York 1975.

[4]J. P. Aubin: Fuzzy Games. Report MRC–1480. Math. Res. Center Univ. of Wisconsin 1975.

[5]J. P. Aubin: Fuzzy core and equilibria of games defined in strategic form. In: Directions in Large–Scale Systems (Y. C. Ho, S. K. Mitter, Eds.). Plenum, New York 1976.

[6]J. P. Aubin: Mathematical Methods of Game and Economic theory. North–Holland, Amsterdam 1979.

[7]J. P. Aubin: Optima and Equilibria. Springer–Verlag, Berlin 1993.

[8]J. F. Baldwin, N. C. F. Guild: A model for multi–criterial decision–making using fuzzy logic. Int. J. on Man–Machine Studies 12 (1978), 87–98.

[9]J. Bezdek, B. Spillman, R. Spillman: Fuzzy measures of uncertainty and consensus for group decision-making. Proceedings of IEEE Conf. on Decision and Control. New Orleans 1977.

[10]O. Bondareva: Jadro igr N lic (in Russian: Core of N-person games). Vestnik Leningradskogo Univ. 13 (1962), 141–142.

[11]D. Butnariu: Fuzzy games. A description of the concept. Fuzzy Sets and Systems 1 (1978), 181–192.

[12]D. Butnariu: Solution concepts for n-persons fuzzy games. In: Advances in Fuzzy Set Theory and Applications. (M. M. Gupta, R. K. Ragade, R. R. Yager, Eds.). North–Holland, Amsterdam 1979.

[13]D. Butnariu: Stability of Shapley value for n-persons fuzzy games. Fuzzy Sets and Systems 7 (1982), 191–207.

[14]D. Butnariu: Values and cores of fuzzy games with infinitely many players. International Journal of Game Theory 16 (1987), 43–68.

[15]D. Butnariu, E. P. Klement: Triangular Norm–Based Measures and Games With Fuzzy Coalitions. Kluwer, Dordrecht 1993.

[16]D. Dubois, H. Prade: Fuzzy numbers: An overview. In: Analysis of Fuzzy Information (J. C. Bezdek, Ed.), CRC–Press, Boca Raton 1988, Vol. 2, 3–39.

[17]J. Kacprzyk: Zbiory rozmyte w analizie systemowej (in Polish: Fuzzy Sets in System Analysis). PWN, Warsaw 1986.

[18]E. K. Kerre, M. Mareš, R. Mesiar: Orderings of generated fuzzy quantities. In: Proceedings of IPMU'98, Paris 1998, La Sorbone 1998, Vol. I, 250–254.
[19]R. D. Luce, H. Raiffa: Games and Decisions. J. Wiley and Sons, London - New York 1957.
[20]M. Mareš: Additivity in general coalition games. Kybernetika 14 (1978), 5, 350–368.
[21]M. Mareš: General coalition games. Kybernetika 14 (1978), 4, 245–260.
[22]M. Mareš: Combinations and transformations of some general coalition games. Kybernetika 17 (1981), 1, 45–61.
[23]M. Mareš: Sufficient conditions for solution existence in general coalition games. Kybernetika 21 (1985), 4, 251–261.
[24]M. Mareš: Fuzzy cooperation without side payments. In: Transactions of the 12th Prague Conference on Information Theory, Statistical Decision Functions and Random Processes. Prague 1994. ÚTIA, Prague 1994, 142–144.
[25]M. Mareš: Computation Over Fuzzy Quantities. CRC–Press, Boca Raton 1994.
[26]M. Mareš: Coalition forming motivated by vague profits. In: Proceedings of the Mathematical Methods in Economy, Symposium, Ostrava 1995. TU of Ostrava 1995, 114–119.
[27]M. Mareš, R. Mesiar: Composition of shape generators. Acta Math. et Inform. Univ. Ostraviensis 4 (1996), 1, 37–45.
[28]M. Mareš: Weak arithmetics of fuzzy numbers. Fuzzy Sets and Systems, 91 (1997), 2, 143–154.
[29]M. Mareš: Fuzzy coalition forming. In: Proceedings of 7th IFSA World Congress, Prague 1997. Academia, Prague 1997. Vol. III, 70–73.
[30]M. Mareš: Verbal quantitative information in economic reality. Acta Oeconomica Pragensia 5 (1997), 1, 121–134.
[31]M. Mareš: Following fuzzy instructions. In: Proceedings of Prague Stochastics'98, Prague 1998, Union of Czech Math. and Phys., Prague 1998, 367–371.
[32]M. Mareš: Additivities in fuzzy coalition games with side-payment. Kybernetika 35 (1999), 2, 149–166.
[33]M. Mareš: Sharing vague profit in fuzzy cooperation. In: Soft Computing in Financial Engineering (J. Kacprzyk, R. Ribeiro, R. R. Yager, H. J. Zimmermann, Eds.). Physica Verlag. Heidelberg 1999, 51–69.
[34]M. Mareš: Superadditivity in fuzzy extensions of coalition games. Tatra Mountains Math. Publ. To appear.
[35]M. Mareš: Fuzzy coalition structures. Fuzzy Sets and Systems 114 (2000), 1, 23–33.
[36]M. Mareš: Superadditivity in general fuzzy coalition games. Kybernetika. To appear.
[37]M. Mareš: Fuzzy Shapley Value: Transactions of IPMU2000, Madrid. CSIC Madrid, Vol. 3, 1368–1372.
[38]M. Mareš, R. Mesiar: Calculation over verbal variables. In: Computing With Words in Information (J. Kacprzyk, L. A. Zadeh, Eds.). Intellingent Systems 1. Physica–Verlag, Heidelberg 1999, 409–427.
[39]M. Mareš, M. Vlach: Concept of linear fuzzy coalitional game. Tatra Mountains Math. Publications. To appear.
[40]M. Mareš, M. Vlach: Linear coalitional games and their fuzzy extensions. Int. J. of Uncertainty, Fuzziness and Knowledge–Based Systems. Submitted.

[41]R. Mesiar: Fundamental triangular norm based tribes and measures. J. Math. Anal. Appl. 177 (1993), 633–640.
[42]R. Mesiar: A note on T-sum of $L - R$ fuzzy numbers. Fuzzy Sets and Systems 79 (1996), 259–261.
[43]I. Nishizaki, M. Sakawa: Fuzzy and Multiobjective Games for Conflict Resolution. Physica–Verlag, Heidelberg. To appear.
[44]J. von Neumann, O. Morgenstern: Theory of Games and Economic Behaviour. Princeton Univ. Press, Princeton 1953.
[45]H. Nurmi: On fuzzy games. In: Proceedings of 3rd European Meeting on Cybernetics Systems Research. Wieden 1976.
[46]J. Rosenmüller: Kooperative Spiele und Märkte. Springer–Verlag, Heidelberg–Berlin 1971.
[47]J. Rosenmüller: Extreme Games and Their Solutions. Springer–Verlag, Berlin–Heidelberg 1977.
[48]J. Rosenmüller: The Theory of Games and Markets. North Holland, Amsterdam 1982.
[49]L. S. Shapley: The value for n-person games. In: Contributions to the Theory of Games II. Annals of Math. Studies 28 (1953), 307–317.
[50]L. S. Shapley: Core of convex games. The RAND Corporation Report, RM–4571–PR, 1965.
[51]L. S. Shapley: On balanced sets and cores. The RAND Corporation Report, RM–4601–PR, 1965.
[52]S. Spillman, J. Bezdek, R. Spillman: Coalition analysis with fuzzy sets. Kybernetes 8 (1979), 203–211.
[53]X. Wang, E. E. Kerre: Reasonable properties for the ordering of fuzzy quantities. Part I, Part II. Fuzzy Sets and Systems. To appear.
[54]L. A. Zadeh: Fuzzy sets. Information and Control 8 (1965), 3, 338–353.
[55]H. J. Zimmermann: Fuzzy Set Theory and Its Applications. Kluwer, Dordrecht 1991.

Index

GPSR Compliance

The European Union's (EU) General Product Safety Regulation (GPSR) is a set of rules that requires consumer products to be safe and our obligations to ensure this.

If you have any concerns about our products, you can contact us on ProductSafety@springernature.com

In case Publisher is established outside the EU, the EU authorized representative is:

Springer Nature Customer Service Center GmbH
Europaplatz 3
69115 Heidelberg, Germany

Batch number: 10725802

Printed by Printforce, the Netherlands